A GAM
Between the Bark Sunbeam and the Ship James Arnold
Painting owned by Mr. Alexander G. Grant

THE YANKEE WHALER

CLIFFORD W. ASHLEY

WITH AN INTRODUCTION BY
ROBERT CUSHMAN MURPHY
AND
A PREFACE TO THE PICTURES BY
ZEPHANIAH W. PEASE

DOVER PUBLICATIONS, INC.
New York

Copyright © 1926 and 1938 by Clifford W. Ashley.
Copyright © renewed 1954, 1966 by Sarah Ashley Delano.
All rights reserved under Pan American and International Copyright Conventions.

Published in Canada by General Publishing Company, Ltd., 30 Lesmill Road, Don Mills, Toronto, Ontario.
Published in the United Kingdom by Constable and Company, Ltd., 3 The Lanchesters, 162–164 Fulham Palace Road, London W6 9ER.

This Dover edition, first published in 1991, is a republication of the work published by Halcyon House, Garden City, New York, in 1942, which was the second edition of the work originally published by Houghton Mifflin Company, Boston, in 1926. A number of the plates have been moved from their original positions, and color illustrations are reproduced here in black-and-white. These alterations have resulted in a reorganization of the list of illustrations on pp. xix–xxiii.

Manufactured in the United States of America
Dover Publications, Inc., 31 East 2nd Street, Mineola, N.Y. 11501

Library of Congress Cataloging-in-Publication Data

Ashley, Clifford W. (Clifford Warren), 1881–1947.
    The Yankee whaler / Clifford W. Ashley ; with an introduction by Robert Cushman Murphy and a preface to the pictures by Zephaniah W. Pease.
      p.   cm.
    Reprint. Originally published: 2nd ed. Garden City, N.Y. : Halcyon, 1942.
    Includes bibliographical references and index.
    ISBN 0-486-26854-3
    1. Whaling.   2. Whales.   3. Whaling—Massachusetts—New Bedford.   I. Title.
SH381.A8   1991   91-18751
639.2′8′09744—dc20   CIP

Inscribed
to the Memory
of
THE YANKEE WHALEMAN

ACKNOWLEDGMENT

IN this paragraph I wish to acknowledge my indebtedness to various people for the assistance I have received in the preparation of this volume, and to express my appreciation of the cheerful willingness with which it has been given: to Frank Wood, F. Gilbert Hinsdale, Martha M. Hinsdale, Philip Sawyer, Edward F. Sanderson, and John Hall Jones, who have allowed me to reproduce in the plates objects from their collections; to George H. Taber, through whose courtesy I am enabled to quote the log of the sloop *Manufactor* of Dartmouth; to Captain William I. Shockley, who has unflaggingly supported his end of a long correspondence, and whose practical knowledge of the Arctic Fishery has cleared many a debated point; to Dr. F. A. Lucas, Dr. Robert Cushman Murphy, and Irving R. Wiles, who have read portions of my manuscript and made suggestions and corrections; to George H. Tripp, of the New Bedford Public Library, who has frequently been called upon for aid and who has never failed to respond; to Arthur C. Watson, whose familiarity with the log-books of the New Bedford Whaling Museum has saved me much research; to Harper and Brothers, the Century Company, and Charles Scribner's Sons, who have permitted the reproduction of copyright material that had previously appeared in their publications; to the Whaling Museum of the Old Dartmouth Historical Society in New Bedford, whose collection of whale gear has been heavily drawn upon for illustration; to my many friends and patrons who have obligingly permitted me to reproduce their paintings; and to a host of anonymous individuals from whom in the course of years I have gleaned many of the facts herein recorded;—to all these my thanks are due, and are here most gratefully tendered.

FOREWORD TO SECOND EDITION

A NUMBER of errors found in THE YANKEE WHALER as it originally appeared have been corrected. There was one of omission, first called to my attention by the reviewers, about which I have since received a number of letters from private sources. This was my failure to mention sea shanties and to give more than cursory attention to the music of the forecastle.

The sea shanty developed spontaneously in the merchant service, where it served a very definite purpose on ships of many nations until donkey engines superseded man-power on deck. On the whaler, however, its employment was intermittent and generally incidental. It never became an important part of the daily life of the whaleman. Once on the whaling ground, during daytime, all unnecessary noise was discouraged. Even orders were given in lowered voices so that whales might not be frightened off. Usually in the dog-watches all hands were permitted to dance, sing, and play whatever musical instruments they had. But there was no occasion for shantying after the lookout had once left the masthead, for ships were then under shortened sail until sun-up. The one purpose of the sea shanty was to lend rhythm to heavy hauling.

Sometimes when getting up anchor, or when cutting-in a whale, there would be a burst of shanty. When homeward bound with a full ship a crew often shantied, but it was then more an expression of their gladness of heart than it was an adjunct to their work. Whalers were small and watches disproportionately large. About forty hands, all told, made up the crew of an average vessel. In consequence working ship was mere child's play beside the heavy grind aboard a more short-handed merchantman. There every pound a man possessed had to be brought to bear on his task.

I would like to claim the shanty for the whaleman, but in all fairness it appears to belong to the merchant service, where it originated, developed, and became obsolete in so short a space of time that the life span of many a sailor bridged the whole period.

Wherever men congregate there is always an attempt to produce harmony of some sort, either vocal or instrumental. Undoubtedly there was music in every whaler's forecastle; an entry in the log of the ship *Ceres* of Wilmington, Delaware, preserved in the Mariners' Museum at Newport News, gives evidence of either a band, an orchestra, or a fife and drum corps aboard ship; but this may be regarded as unusual.

"Jul. 4, 1834.—In evening mustered up our drum, fifes, flutes and fiddles to celebrate Independence."

The *Ceres* voyage proved a failure and it seems very possible that the crew had been chosen with the wrong talents in view for a successful business venture.

One question of fact has been raised—Mr. Brindley, who reviewed THE YANKEE WHALER in the *Mariner's Mirror* (October, 1928), doubts my attribution of an American origin to scrimshaw. So far as I know this has not been questioned before; it has always been tacitly accepted in accordance with the published records.

Since scrimshaw is an important folk art, perhaps the only one we can regard as peculiarly American, his opinion should not be allowed to go unanswered; particularly as the *Mariner's Mirror*, in which it appeared, is a serious publication devoted to maritime research, which ordinarily carries considerable authority. It is well known over here as the official organ of the British "Society for Nautical Research," whose activities in saving Nelson's flagship *Victory* have earned the respect and gratitude of all sea-minded Americans.

In justice to Mr. Brindley, I do not think he would have voiced this opinion if he had been on familiar ground: he gives as authority for his criticism the catalogue of the Hull Fisheries Museum. His paragraph dealing with scrimshaw runs as follows:

"*We feel that he (Ashley) makes too wide a claim in saying that 'it (scrimshaw) is the only important indigenous folk art, except that of the Indians, we have ever had in America.' Does not Mr. Ashley think it possible that British or Dutch whalemen on the early Greenland Fishery may have lessened the tedious hours by scratching designs on teeth or fragments of bone? ... The current catalogue of the Hull Fisheries and Shipping Museum states that some of the scrimshaw therein is work of two centuries ago. It is dangerous to attribute any simple and widely diffused art to a particular people when it has been practiced for more than a hundred years.*"

Of course indigenous is not an exclusive term, and a thing actually may be indigenous in several places, but that being the case I was perhaps at fault in not making the unequivocal statement that scrimshaw, having originated in American ships, was restricted to them until the time of the American Revolution.

The Greenland whale being entirely toothless, I do not take seriously the suggestion that Greenland whalemen lessened any of their tedious hours by scratching designs on teeth. The Sperm Whale is the *only* whale that has teeth (except the dolphin family, and one dwarf species) and the Sperm Whale lives only in tropical and sub-tropical waters. It is the Sperm Whale alone that provides the usual ma-

FOREWORD TO SECOND EDITION

terials of which scrimshaw is made; the tooth is preferred, but for large pieces of scrimshaw bone is used, and only the pan-bone of the Sperm Whale's jaw is considered fit for the purpose. The rest of his skeleton and the bone of other whales is much too porous.

There were no British Sperm Whalers until the beginning of the American Revolution, at which time captured American whalers were sent to sea under the British flag. (THE YANKEE WHALER, pp. 25-26.)

The Sperm Whale Fishery was instituted by New England Whalemen in the beginning of the eighteenth century and was never prosecuted by any other people until the British fishery participated.

For the sake of clarity I shall use the terms "British" and "American" as if they referred to two different fisheries, although the American fleet was a part of the British fishery until July 4, 1776.

Since THE YANKEE WHALER was first published, a number of early dated pieces of scrimshaw have been brought to my attention. Among them one of the best authenticated is a delicately carved and decorated pepper shaker of sperm ivory, dated 1773, now in the Hinsdale collection. It was made on a New Bedford ship by William Carsley, Sr., father of William Carsley, the inventor and patentee of the so-called "Carsley Iron," who was grandfather of Edward R. Cole, one of the last two whalecraft manufacturers in the New Bedford district. Mr. Cole is co-donor of the Whalecraft Exhibit in the New Bedford Whaling Museum, and it was from him this example was obtained. The pepper shaker antedates by several years British participation in the Sperm Whale Fishery.

When I visited the Hull Museum in the spring of 1928, I was particularly interested in the Scrimshaw Exhibit. There were two items, at that time, included among the scrimshaw, that are dated earlier than the commencement of British Sperm Whaling. These two pieces are listed as scrimshaw in the catalogue of 1928, and are undoubtedly the "work of two centuries ago" referred to by Mr. Brindley. They are snuffboxes made of walrus ivory and wood (teakwood being a material of one of them). Both are similar to commercial snuffboxes to be seen in various snuffbox collections, and are distinct in character from the typical scrimshaw of more recent date which comprises the rest of the exhibit.

Walrus ivory was a regular commodity of trade as early as 1617 (see "Purchas His Pilgrimage," 1617, p. 931). It was the best material then obtainable for making false teeth, and was also commonly used in the cutlery trade. It is harder than elephant ivory and brought a higher price; hardly a material to be given out in the forecastle.

Although separated by about half a century, the two boxes bear the same legend, a common one on commercial snuffboxes, and the spelling is identical: "If you love

FOREWORD TO SECOND EDITION

mee, lend *mee* not." The fact of the English motto would seem sufficient to preclude a Dutch origin, unless it were a commercial one.

The date of the earlier box is 1665. Being at war with England, the Dutch sent out no whaleships in 1665. Between 1660 and 1670 the British averaged less than one whaler at sea per season, so it is possible that there was a ship at sea in 1665 on which a snuffbox could have been made. Between 1679 and 1694 there were no British whaleships at sea, although huge bounties were being offered by a government anxious to break the Dutch Greenland monopoly. In 1693 "The Company of Merchants of London trading in Greenland" was incorporated for a term of fourteen years with a paid-in subscription of £40,000, later increased to £80,000. Its funds were exhausted prior to the expiration of its charter, and some time previous to 1707 the British Fishery was again abandoned, not to be revived until the South Seas Company secured a favorable grant and sent twelve new vessels to Greenland in 1725. These dates may all be found in Macpherson's "Annals of Commerce," and Scoresby's "Arctic Regions and Northern Whale Fishery," two outstanding British authorities, the latter probably being "the" authority on the British Greenland Fishery.

The second snuffbox is dated 1712. In that year there were no British whaleships, there had been none for over five years, and thirteen years was to elapse before another one put to sea. There appears to be nothing further in the Hull Museum catalogue that could be taken as evidence against the American origin of scrimshaw.

CLIFFORD W. ASHLEY

WESTPORT, MASS.
February, 1938

FOREWORD

WHEN THE YANKEE WHALER was first planned, I intended to use as the basis for my text the account of my whaling voyage in the bark *Sunbeam*, originally published in *Harper's Magazine* some twenty years ago under the title "The Blubber-Hunters." I proposed to rewrite this entirely, elaborating upon the original and adding to it certain historical facts and descriptive matter pertaining to the methods and paraphernalia of the fishery.

But when confronted with the task of revision, I was soon persuaded that this boyhood account of mine, although slight, was nevertheless a first-hand document; a contemporary record of one of the last voyages ever made by a New Bedford square-rigger; and for that reason it might well be left unchanged.

So except for the removal of some errors, and the return of a few paragraphs that were in the original manuscript, the story stands; stands as it first appeared, at a time when there were still half a hundred whalers sailing out of New Bedford, and when many near-sighted people of the old town still believed that the whale fishery was too big a thing ever to perish.

In the supplementary chapters which have been written to contain the additional matter wanted for this volume, I have been more concerned with the outward semblance of the things that I have known than I have been with their economic and historical significance. Other and better historians—Scoresby, Macy, Scammon, and Starbuck—have told the story of the rise and fall of the whaling industry, so I have not attempted to give either a complete or a connected historical account. It will never be possible again to study the old whale fishery at first hand, so I have not hesitated to tell my own experiences and to describe at length the things that have interested me—the development of the whaler from the small sloop, the evolution of the whaleboat from the Indian canoe, the use of the whaleman's implements and tools; the story of the whaleman himself, the manner of his fishing, and the *scrimshaw* of his leisure hours. To these I have added certain observations regarding the natural history of the whale.

There will be some account of the methods of early fisheries, but whaling has changed surprisingly little through the centuries. Scoresby in 1820 wrote, "the way in which the whale is pursued and killed is pretty nearly the same at this time as it was a hundred years ago." This statement could have been made with equal truth either one hundred years earlier, in 1720, or one hundred years later, in 1920.

We took our whales on the *Sunbeam* with a hand harpoon in a boat pulled by six men, and, except that the head of the harpoon was of different shape and the boat carried a bigger sail, there was no fundamental change from the days of Scoresby, or even of Edge, two hundred years earlier. So far as I know, I sailed upon the last merchant ship that ever carried hemp standing rigging. The *Sunbeam's* jackstays were wooden battens and she crossed a spritsail yard. Without doubt she was a crawling, creaking anachronism, but to me she was the most beautiful thing that ever sailed the seas.

The scanty account to be given of early foreign whaling is drawn largely from Scoresby's admirable work, "An Account of the Arctic Regions and Northern Whale Fishery," London, 1820.

For data on the early American whale fishery, I have depended principally on the files of the *New Bedford Mercury*, the New Bedford "Whalemen's Shipping List and Merchants' Transcript," Scammon's "Marine Mammalia and American Whale Fishery," Macy's "History of Nantucket," Ricketson's "History of New Bedford," Alexander Starbuck's "History of the American Whale Fishery," and on old whalers' log-books preserved in the Public Library and the Whaling Museum of New Bedford. In a bibliography at the back of this volume is a list of the books that I have had constantly by me while preparing this work. Generally when I have quoted I have named my authority, but frequently when I have quoted from notes, this has not been possible. These notes were made over a considerable period of years, and often I failed to make attributions. Sometimes it has been necessary to quote from memory, but I have not consciously indulged in unverified statement.

Few writers of either whaling fiction or history have had personal experience on the deck of a whaler. Whaling was such a highly specialized industry, and the pursuit of it was so elaborately technical, that inevitably they have drifted into many errors, which now threaten by frequent repetition to become standardized and accepted as facts. For this reason, I have elaborated upon a number of points that may have little interest for the layman. But I trust that there is enough drama in the main facts of the American whale fishery to carry the reader through these doldrums.

The Yankee whaleman has written for himself a glorious page in American history. His courage and hardihood are so well attested that his reputation should be secure. But there has been a tendency among writers of fiction to discredit him. Certain widely circulated books have painted him in false and lurid colors. So long as this was confined to fiction, no great harm was done. But when fiction began to masquerade as fact, and sciolists solemnly borrowed from fiction, labeling their wares "history," it was time for some one to come to the rescue of the whalesman's tarnished reputation. I have endeavored to correct some of these fallacies. Three institutions, the New Bedford Whaling Museum, the New Bedford Public Library, and the newly founded Nantucket Whaling Museum, are now doing excellent work toward the preservation of the history of the whaleman's achievement.

The plates in this volume reproduce, with few exceptions, all the pictures I have made that concern whaling. The first ten years of my illustrative work are responsible for the drawings contained in "The Blubber-Hunters," originally published in *Harper's Magazine*, and for the set of "Subject Pictures," which were first reproduced in *Century* and *Scribner's Magazines*.

About 1914 I began to paint the few remaining whaleships at their wharfs in New Bedford, for it seemed to me that a record should be made. My task came to a close last summer when the *Charles W. Morgan* was towed out of the harbor and there were no whalers left to paint.

Among the illustrations are included a number of photographs, mainly of whalecraft, whalegear, and scrimshaw, and several plans and diagrams designed to supplement the matter contained in the paintings.

CONTENTS

INTRODUCTION, *by Robert Cushman Murphy, D.Sc.*	xxv
A PREFACE TO THE PICTURES, *by Zephaniah W. Pease*	xxvii
I. THE BLUBBER-HUNTERS	1
II. THE BLUBBER-HUNTERS, *Concluded*	11
III. THE GREENLAND FISHERY	23
IV. CAPE COD, LONG ISLAND, AND NANTUCKET	28
V. NEW BEDFORD	34
VI. THE WHALER	44
VII. THE WHALEBOAT	57
VIII. THE WHALE	65
IX. GEAR AND CRAFT	85
X. THE WHALEMAN	99
XI. SCRIMSHAW	111
XII. THE LAST DAYS OF WHALING	117
EPILOGUE	121
A GLOSSARY OF WHALING TERMS	123
A LIST OF BOOKS CONCERNING WHALES AND WHALING	147
INDEX	151

ILLUSTRATIONS

A GAM BETWEEN THE BARK SUNBEAM AND THE SHIP JAMES ARNOLD *frontispiece*

Headpiece: UNDER WEIGH   I

Following page 10
HARPOONING A PORPOISE
CUTTING-IN: AN 85-BARREL WHALE ALONGSIDE THE SUNBEAM, 1904
BAILING THE CASE
FAST BOAT

Headpiece: THE MONKEY ROPE   11
Headpiece: A GREENLAND WHALER   23
Headpiece: THE SHORE LOOKOUT   28
Headpiece: QUARTERDECK OF THE BARK SUNBEAM   34

Following page 34
THE WANDERER IN PORT, SEPTEMBER, 1920
SHIP NIGER, FULL AND BY
BARK STAFFORD OUTWARD BOUND
DRYING SAIL: BARK WANDERER, AUGUST, 1920

Headpiece: EVOLUTION OF THE WHALER   44
Headpiece: BOATS AWAY   57

Following page 64
THE CHASE OF THE BOWHEAD WHALE
SAILING DAY
THE FLURRY
LANCING A SPERM WHALE
SAMPLING OIL AND DRYING SAILS
COOPERING CASKS
CLOSE-HAULED: THE BARK ANDREW HICKS
JOURNEY'S END: THE WRECK OF THE WANDERER

ILLUSTRATIONS

Headpiece: BARK OSCEOLA 3RD ATTACKED BY A SPERM WHALE 65

Following page 84

STOVE BOAT
THE LAST WHALER: THE CHARLES W. MORGAN TIED UP AT
    NEW BEDFORD
FITTING OUT THE SUNBEAM, 1904
FITTING OUT BARK ANDREW HICKS
THE GREENLAND FISHERY
THE SAIL LOFT
CAULKING A WHALER
"THE FOOT OF THE STREET"
BARK SUNBEAM BECALMED
BOTTLE-NOSE PORPOISES
THE DAY BEFORE SAILING
THE WANDERER TIED UP
GATHERING FOG
THE GREYHOUND OUTFITTING
OUTFITTING THE CHARLES W. MORGAN, 1916
THE BARK COMMODORE MORRIS OFF DUMPLING LIGHT
THE STERN OF THE WANDERER
A WHALER AT UNION WHARF
THE WANDERER, FROM FISH ISLAND

Headpiece: GRINDSTONE AND COOPER'S DEVIL 85

Following page 98

SUNBEAM LOWERING BOATS
STRIPPING THE WANDERER, 1923
WHALER AND BUMBOATS AT BRAVA
THE GREYHOUND IN PORT, 1918
BEFORE THE MAST: A FORWARD DECK VIEW OF THE
    CHARLES W. MORGAN
THE GREYHOUND: WITH SAILS BENT PREPARATORY TO SAILING
GRAY FOG
THE WANDERER WITH SAILS CLEWED
SPERM WHALING: SKETCH FOR A LUNETTE
BOWHEAD WHALING: SKETCH FOR A LUNETTE
WIND FROM THE NORTHWEST

ILLUSTRATIONS xxi

THE CAPTAIN: THE SUNBEAM'S AFTER CABIN
THE GREYHOUND AND OTHER CRAFT
WRECK OF THE WANDERER AT CUTTYHUNK, AUGUST 26, 1924
THE SUNBEAM CRUISING
SCHOONER A. E. WHYLAND AND MERCHANT BARKENTINE SAVOIE
HARBOR MIST: THE CHARLES W. MORGAN AT KELLEY'S WHARF, FAIRHAVEN
ANCIENT CRAFT, 1914
THE BARK SWALLOW
THE WANDERER IN WINTER QUARTERS
THE CHARLES W. MORGAN AT FAIRHAVEN
A WHALESHIP ON THE MARINE RAILWAY AT FAIRHAVEN
A WHALER OUTWARD BOUND
THE COOPER SHOP: FIRING THE CASK
THROUGH THE STRAITS
FAIRHAVEN
THE SUNBEAM IN A FOG

Headpiece: TOWING THE CATCH TO SHIP 99

Following page 110

OLD NEW BEDFORD
SAILING OF THE DESDEMONA
OCTAGON BUILDING AT FAIRHAVEN
COOPERING CASKS FOR THE GREYHOUND
THE SWALLOW INSHORE OFF CUTTYHUNK
NEW BEDFORD HARBOR: THREE-MASTED WHALING SCHOONER MYSTIC DRYING SAILS
BLUFF BOWS: THE CHARLES W. MORGAN
THE SAILMAKER
THE CHARLES W. MORGAN FROM THE STATE PIER
THE CHARLES W. MORGAN AT SEA
THE CHARLES W. MORGAN AT ROUND HILLS
"A DEAD WHALE OR A STOVE BOAT!" THE HERMAPHRODITE BRIG DAISY, HOVE TO WITH BOATS DOWN
NEW BEDFORD WATERFRONT, SHOWING THE MERCHANT BARK CHARLES C. RICE AND THE HERMAPHRODITE BRIG HARRY SMITH
THE SUNBEAM WEARING SHIP IN A FOG

ILLUSTRATIONS

THE OLD AND THE NEW
GETTING FAST
THE GREYHOUND AT MERRILL'S WHARF
BARK WANDERER CLOSE-HAULED
THE SUSIE PRESCOTT AND THE CHARLES W. MORGAN
THE SUNBEAM HAULED OUT
THE SHIP CARPENTER: AT THE SUNBEAM'S CUTWATER
CAULKING THE SUNBEAM
MOUNTING AN IRON
"THERE SHE BLOWS!"
BOATS AWAY FOR WHALES
SUNBEAM REACHING
"A NANTUCKET SLEIGH-RIDE"
HOISTING BOATS: THE SUNBEAM, 1904
THE FIRST BLANKET PIECE
MINCING
BOILING: THE SUNBEAM'S TRY-WORKS

Headpiece: THE FORECASTLE   111

Following page 116

SCRIMSHAW WHALE'S TEETH
SCRIMSHAW: SPERM WHALE'S TEETH
SCRIMSHAW: TWO SPERM WHALE'S TEETH
SCRIMSHAW: COCOANUT DIPPERS
SCRIMSHAW: MOPS, FORK, ROLLING-PIN, ETC.
SCRIMSHAW: JAGGING-WHEEL, SALT-SPOON, ETC.
SCRIMSHAW: BUSKS
SCRIMSHAW: CANES
SCRIMSHAW: DITTY-BOX
SCRIMSHAW: A WHALING SCENE ETCHED ON PAN BONE
THE SECOND MATE'S DITTY-BAG
SCRIMSHAW: KNIFE-BOX, SILK SWIFT, MORTAR AND PESTLE,
    SALT-SHAKER, AND DITTY-BOX
MODELS OF SIX WHALES
SAILORS' KNOT WORK
SIX SAILORS' CHESTS SHOWING ORNAMENTAL KNOTWORK AND
    NEEDLEWORK BECKETS

ILLUSTRATIONS xxiii

FIGUREHEAD OF THE MARTHA, OF NEW BEDFORD
BILLETHEAD FROM THE BARK ROUSSEAU
FIGUREHEAD OF THE MERMAID
FIGUREHEAD OF THE BARTHOLOMEW GOSNOLD
FOUR WHALERS' STERNBOARDS
CARVED AND PAINTED STERNBOARD OF THE SHIP MARY AND SUSAN, OF STONINGTON
SIGNAL BOARD OF THE NORTHERN LIGHT
WHALE HARPOONS
WHALE HARPOONS: A GROUP OF EXPERIMENTAL IRONS
WHALE HARPOONS
BOAT GEAR AND CRAFT
WHALE GEAR IN THE COLLECTION OF THE NEW BEDFORD WHALING MUSEUM
WHALE GEAR: BLUBBER-PIKE, BLUBBER-FORKS, ETC.
CAULKER'S TOOLS AND RIGGER'S BELT
GREENER BOW GUN
WEIGHING WHALEBONE

HEADPIECE: FURLING SAIL   117
MODELS OF A RIGHT WHALE AND A SPERM WHALE   122

Following page 122

A SIXTEEN-FOOT SPERM WHALE'S JAW
WHALEBONE: BALEEN FROM THE MOUTH OF THE BOWHEAD WHALE
FLUKES OF A SEVENTY-FIVE BARREL SPERM WHALE
WHALING COSTUMES FROM THE SUNBEAM, 1904
THE CAPTAINS OF THE "STONE FLEET"
MODEL OF A WHALEBOAT
PLAN OF AN ENGLISH SEVEN-MAN BOAT THIRTY-TWO AND ONE HALF FEET LONG
PLAN OF A THIRTY-FOOT NEW BEDFORD WHALEBOAT
MODEL OF THE LAGODA
LINES "TAKEN OFF" THE ORIGINAL BUILDER'S BLOCK MODEL OF THE BARK SUNBEAM
SAIL-PLAN MADE IN HART'S SAIL LOFT, NEW BEDFORD, IN 1883, FOR THE BARK ALICE KNOWLES
DECK PLAN AND FORE AND AFT SECTION OF THE BARK ALICE KNOWLES

INTRODUCTION

TO those of us who are familiar with the literature of Yankee whaling, Clifford W. Ashley's book comes as the swan song for which we had hardly dared to hope. Since the days of "Moby Dick" and before, the gamut of the ever-fascinating subject has been run, and the romantic, the historical, the descriptive and statistical, the geographic and sociological aspects have received well-nigh uninterrupted treatment by many authors. The product may truly be said to have ranged from the sublime to the ridiculous, but, with a few notable exceptions, angels have not trod the field into which so many others have rushed.

The reason for the dearth of authentic records of American whaling from the pens of participants is a simple one. The bulk of those who were equipped to recount the tale were too close to their subject to acquire perspective. They recognized nothing remarkable in the intimate details of their daily tasks, and at best they sensed dramatic interest only in the spectacular elements of combat and capture. The New Bedford skipper with whom the writer of these lines sailed in 1912-13 had been steeped for seventy years in the lively environment of whaling, yet he was indisposed to dole out the fragments of his Odyssey. When it came to sea-elephant hunting, a relatively novel pursuit to the New Bedford imagination, he was ready enough to spin yarns and even to burst into print; but, as for whaling, it was but the commonplace of the world in which he moved.

Moreover, in latter decades, as contrasted with the Golden Age, whaling had for the most part been taken over by men who lacked the ability to exploit their own experience even had they so wished. Several of the most popular historians were by no means fighting whalemen, but rather sea-cooks and ship-keepers whose lots chanced to be cast with a whaler instead of a merchantman. The remaining authors have either considered whaling as a fishing industry, or else they have caught but the reflection of high romance by cajoling old boat-headers into garrulity, by drawing out the disgruntlements of forecastle hands who had spent their winnings, by lifting from log-books the raw pabulum which they were not wholly capable of assimilating, or by culling their tales from the scattered literature to which age had brought oblivion.

But now we have the story of the Yankee Whaler from a man who has both delved into the past and lived through the action—from the student, the artist, and the whaleman all in one. Born in New Bedford, of blubber-hunting ancestry, Clifford

W. Ashley was fitted by circumstance and temperament to be supremely happy in the ambience of the whaling tradition. He is old enough to have seen a score of square-riggers fitting out at one time, and young enough to have suffered poignant regret when the *Wanderer* piled on the rocks of Cuttyhunk and the end had come. All his life he has been absorbed in tracing the evolution of rig and gear and handicraft as related to whaling. Endowed with more than his fair share of shrewd New England wit, keenly alive to the power and picturesqueness of the whaling unit, the thrill of portless roaming, the bloody and desperate quest — what are we not justified in expecting when Ashley packs his chest and ships aboard the *Sunbeam*?

He went whaling prepared as no man had been before him, and he went with an eager and an open mind. He describes as no other author has done the multifarious duties of the cooper, the whalecraftsman, the rigger, the subtleties of boat gear and of stowage, the homely whims and prejudices of the most conservative and most practical of men. He nails to the mast, once and for all, a multiplicity of lies and misapprehensions which have crept into the record of whaling. He sees clearly the worst side as well as the best, but he has returned unsoured from his voyaging, and now lays his booty before us.

ROBERT CUSHMAN MURPHY

AMERICAN MUSEUM OF NATURAL HISTORY
  NEW YORK, *April* 20, 1926

A PREFACE TO THE PICTURES

IN the last hours of the expiring whaling industry the fear came upon those who had been in touch with the old days of New Bedford's glory on the seas, that in a few years there would be left but a meager record and reminder of the Golden Age of our history when we prospered by the strange and fascinating vocation of whaling. A new industry had transformed the town and its people. "I did not see the town of my dreams," said the visitor. "The Spouter Inn was not, albeit a man did show me the long lance, 'now widely elbowed,' with which ' Nathan Swain did kill fifteen whales between a sunrise and a sunset.' Queequeg and Dago and Tashtego I missed. But I saw such a town as Melville saw not, when he wrote, 'Nowhere in all America will you find such patrician-like houses, parks and gardens more opulent than in New Bedford, and all these brave houses and flowery gardens came from the Atlantic and Pacific oceans. One and all they were harpooned and dragged up thither from the bottom of the sea.'"

As the whaling industry waned, the sons and daughters of the great race of whalemen came to the realization that the city was lacking in tangible things to preserve the record of the immortal days. Whaling had inspired one literary classic, "Moby Dick," written by Herman Melville. Otherwise the literature was largely limited to log-books and histories in which great events were reduced to unimaginative records and the merest commonplace.

The great subject of whaling had not fared so well in art as it had in literature. The artists who were contemporary with the palmy days of whaling painted for patrons who had very definite ideas about what they wanted and set up a standard to which the painters of the period adapted themselves. The demand of the time was for portraits of ships. The whaler must be broadside, with her hull drawn to a scale and every rope of the rigging must show. These ship portraits were likenesses, maybe, and demonstrated skill in draughtsmanship — but they were not art.

But at the last moment there appeared a painter, Clifford W. Ashley, whose imagination as a boy had been stirred by the hours he spent on the wharves where lay the ships, and where loitered the whalemen who found their joy and content in the surroundings on shipboard where they had met great adventures. The ships fascinated the boy, and in the early days of his work as an artist they became the subject he set apart for himself. After a period under the instruction and influence of Howard Pyle, the young artist decided to make a whaling voyage in order to learn the

A PREFACE TO THE PICTURES

technique of whaling and he secured a commission from *Harper's Magazine* to write and illustrate a story. "The Blubber-Hunters" was the result, and incidentally it may be said that Ashley is the only writer in this generation of a whaling story in which the speech of whalemen is reproduced as New Bedford people have heard it.

A whaleship, with her wooden davits, the peculiar construction of her after house, her rig, and the numerous whaleboats of distinctive type, has a character all her own. Ashley caught it in these first illustrations as no other had done. He saw the ships under conditions in which other artists had not seen them. We have one in mind where the boats are lowered and the ship is lifted above them on a sea, which brings the vessel's stern into relief like the stern of an ancient caravel, a resemblance which has been remarked by those who have seen whalers from this unusual point of view.

Twenty years have passed since Ashley made that group of illustrations and he has been painting the old wharves and whaleships of New Bedford during that time.

Hundreds of painters have been attracted by the subject and have painted the whaleships at our wharves. But they found the spirit of these ships as elusive as "the primrose by the river's brim." New Bedford accepts Ashley as the master who paints whaleships with authority and suggestion. There are colors — perhaps a touch of green, the colors we see in the burning driftwood — which the hull of a whaler takes on at times, after her copper has been acted upon by the seas and the salts have saturated the hull to indescribable iridescence, which he alone has caught. The whaleships which he paints are full of subtlety. One appreciates the stories they have lived. Ruskin has said, "If it is the love of that which your work represents — if being a landscape painter it is love of hills and trees that moves you — if being a figure painter it is the love of beauty and human soul that moves you — if being a flower or animal painter it is love and wonder and delight in petal and limb that move you — then the spirit is upon you and the earth is yours and the fullness thereof." And the spirit is upon this painter, moved by the love of the sea and the ships.

We sometimes wonder if those who have seen these ship and wharf paintings in galleries away from New Bedford and the surroundings where they must be best understood and appreciated, can realize the romance of the story they tell and the fidelity with which it is suggested. Here in New Bedford these pictures of the last of the fleet to survive the perils of the sea and changed conditions on land pass through the eyes, where the obvious stops, and touch and stir the imagination and the heart.

Zephaniah W. Pease

New Bedford
May 12, 1926

THE YANKEE WHALER

THE YANKEE WHALER

CHAPTER I

THE BLUBBER-HUNTERS

WHALING, of all our early industries, has come down to us to-day the least altered in the lapse of years, the least affected by changed conditions, the least trammeled by modern appliances. Of all pursuits, it has preserved to the greatest degree its original picturesqueness. Modern methods have been applied only to the off-shore fishery; deep-sea whaling, Sperm whaling, differs scarcely at all from the whaling of a century ago.

In August of 1904 I shipped from New Bedford on the bark *Sunbeam*, bound on a Sperm whaling voyage to the West Coast of Africa and Crozet Island Grounds.

The *Sunbeam* is a bark of two hundred and fifty-five tons. She was launched at Mattapoisett before England had laid the keel of the last of her "wooden walls."[1] In 1904 she was the only square-rigger being fitted. Battered and weather-beaten, she lay at her berth, partially dismantled, a swarm of workers patching and caulking her sides.

After her topsides had been repaired she was hauled out on the railway over on the Fairhaven shore. Here her bottom was overhauled and recoppered. Though fifty

[1] *Sunbeam* was wrecked off Sapelo Island, Georgia, 1911.

years old, her keel was straight as a gun-barrel. The *Sunbeam's* sheathing had not been off in fifteen years. Many of her planks were rotten, and in one place a stone, which had got in while she was building, had washed around next her keel, and worn through nearly four inches of planking.

In 1854 one hundred and thirteen whalers sailed from New Bedford; fifty years later, five. That which could not be effected by the capture of thirty-four vessels by the *Shenandoah*, the sinking of thirty-nine in the Stone Fleet of Charleston Harbor, the abandonment in two seasons of fifty-four in the Arctic, and other catastrophes equally destructive if less spectacular, has been accomplished by petroleum. Whaling to-day may be reckoned a dead industry — not that it is extinct, but because it can never recover.

There was some difficulty attendant on my arrangement for a berth on the *Sunbeam*; a whaler is often so full-handed that a bunk must be used watch and watch about, and seldom if ever is there a spare one. The owners finally made a place for me in the steerage along with the boat-steerers, and the captain proffered the freedom of his cabin, once we were under way, and suggested that I arrange with the cooper to set up a bunk for me in his stateroom.

I found Cooper in a Water Street grog-shop where he had already reached a proper sailing-day condition. After a little difficulty I secured a word with him in private. He seemed very much interested in the purpose of my voyage, hinted darkly of literary efforts of his own, as yet unpublished, and suggested collaboration.

The matter of the berth was easily settled. That same afternoon I signed the ship's articles.

The day before we were to clear, Friday, I put my chest on board. My bunk was made up, and the usual calico curtain strung before it. But Saturday brought a gale from the southward, and by noon it was decided to postpone the sailing till Sunday. The *Sunbeam* had put into the stream overnight, and was anchored near the Fairhaven shore, with a part of the crew on board.

The agents of the *Sunbeam*, J. and W. R. Wing, are the oldest firm of ship-owners in the whaling business, having, under the present name and heads, controlled vessels since 1856.[1] The firm's office is in the rear of their clothing and outfitting establishment.

Sunday morning, long before the church-bells rang, we were gathering in the darkened front of the store. I had stopped at the post-office for my last mail, and as I stepped out into the bright sunlit street, a couple of sailors lumbered hastily by and dodged around the corner. As they were vanishing, one of the owners appeared in the street, gazing up and down its length, vainly seeking a glimpse of the runaways. When he saw me he hailed cheerfully. From the alley from which he had emerged a

---

[1] The last Wing ship was sold in 1916.

series of derisive hoots followed him, then a wagon-load of seamen appeared, being trundled off to the river. Swaying and pitching as the cart jolted over the cobbles, they boisterously spoke each passer-by, making the street hideous with their yells. Before I entered the store I saw them, one by one, dropping off over the tailboard, oblivious to the protests of the unfortunate drygoods clerk who was responsible for their delivery.

The front shop was crowded and noisy, but the real hubbub was in a small back room. Here the sailors, howling and pounding, were locked up when caught, and held till the return of the wagon to take them to the waterfront. Word was received that the mate refused to go on board till he had partaken of his Sunday dinner. On various pretexts others sought to get off for a while longer — one had forgotten to bid his mother good-bye; another had left home without an overcoat! The clerks rushed frantically about. Each man had to be rounded up — not once, but half a dozen times.

The morning dragged out toward noon. A carriage had been sent for the mate; the little back room was emptied. Cooper was sitting on the edge of a black-draped counter, and a clerk was vainly trying to induce him to go on board. Smilingly the cooper doomed the man to eternal perdition; then picturesquely started in to abuse his ancestors. The disgusted clerk gave up the job; and Cooper sat on, furtively keeping his eye on the ship's chronometer, knowing full well that of all things this would be the last to be taken on board.

We were on the tug, watching the crew tumbling into the pilot-boat, when a commotion at the far end of the wharf attracted our attention, and a clerk hove into view, puffing like a grampus, dragging behind him at the end of a string one of the boat-steerers. As they drew near, we could see that the string was attached to the much-stretched neck of a tan-colored pup. The owner, holding the unlucky mongrel tightly to his breast, struggled to keep up, but the pace was far too stiff for him. In a state of complete exhaustion, the three made the sloop just as she was casting off.

We followed the pilot-boat out across the Acushnet to where the *Sunbeam* lay, redolent in her newly acquired paint, spick and span from sprit to taffrail. Under her spare boats hung all of a hundred fresh green cabbages, the deck crate was filled with potatoes, and a quarter of beef was suspended from the skids. The hatches were buried completely under a heap of mattresses and baggage. We clambered up the man-ropes, and immediately all was confusion. The green hands, resplendent in new suits of dungaree, were falling over one another in their efforts to execute orders. Without loss of time, the hawser was passed from the tug and the command given to weigh anchor. Amid the clicking of pawls and the groaning of the windlass we got under weigh, and on the outgoing tide we were towed out of the harbor. There were over thirty visitors with us for the trip down the bay, friends and relatives of the captain, crew, and owners, who would return to town with the pilot-boat.

Early in the afternoon a picnic lunch was served to all hands, of cold corned beef and pilot-biscuit. Then the tug left us, and with a creaking of blocks and a hollow flapping, the foretopsails went up, then the jibs, spanker, and maintopsails. We were out to sea now, where we could feel the long regular heave of the ocean, and so we sailed for a couple of hours longer, till the pilot-sloop *Theresa* overhauled us. Our foreyards were backed and the ship hove to. The *Theresa* put up into the wind, and lay a little off our starboard quarter. The davit-tackle of the starboard boat creaked. There was a faint splash, and the friends of the crew were hurried away. The picnic aspect was gone; in its place lurked the emotion of a long parting. Soon the boat came for the second and last load. The owners and their friends, the captain's friends, and the pilot went over the side and were rowed out to the *Theresa*. The crew pulled jerkily and unevenly; it was a far cry to the long whippy stroke of the later season.

And now they rested on their oars, and some one stood up in the stern-sheets, his voice sounding strangely remote from across the water; "A short and greasy voyage!" he called, and the boat and the sloop gave us three rousing cheers. Then we turned to the open sea.

The crew gathered in a silent group at the forecastle, and watched the narrow strip of headland fade slowly away. After awhile the breeze died down, and we drifted with the ebb of the tide back toward Mishaum buoy. Toward nightfall Cooper screwed up the deadlights; and later, the wind freshening from a new quarter, the last vestige of land quickly dropped from our horizon.

The steerage that night was not an inviting place in which to sleep. On a clutter of chests and dunnage the boat-steerers sprawled, drinking, wrangling, smoking. Some had turned in dressed as they had come on board, togged out in all their petty shore finery, and now huddled in inert, lifeless heaps, or half hanging from their berths, with swollen necks and puffed and livid features.

The floor was littered with rubbish, the walls hung deep with clothing; squalid, congested, filthy; even the glamour of novelty could not disguise the wretchedness of the scene. The floor was wet and slippery, the air smoky and foul; often a bottle was dropped in the passing, or an empty one was smashed to the floor. Through it all was an undertone of water bubbling at the ports and a rustle of oilskins swinging to and fro like pendulums from their hooks on the bulkhead. Roaches scurried about the walls. A chimneyless whale-oil lamp guttered in the draft from the booby-hatch and cast a fitful light over the jumble of forms sitting on the chests beneath. Occasionally there was a trampling of feet overhead, and an order was hoarsely shouted.

THE BLUBBER-HUNTERS

The ship rolled gently through the oily seas, the wind hummed drearily through the tautened rigging.

All hands were called aft in the morning immediately after breakfast, and pacing the deck with hands in his pockets, Captain Higgins gave utterance to various sentiments appropriate to the start of a long sea-voyage.

"Just remember, I'm boss on this ship. When you get an order, jump. If I catch any one of you wasting grub, I'll put him on bread and water for a month and dock the rations of the whole watch. You greenies have got just a week to box the compass and learn the ropes; after that, no watch below till you do. Let every man work for the ship; I don't mind a little healthy competition between the boats, but if any dirty work goes on, I'll break the rascal who does it. We've got to work together — see? Now go ahead and pick your watches." Straightway the crew was told off into two lots by the first and second mates, and the starboard watch was sent below.

Of our quota of thirty-nine all told, only eight, including myself, were born American; Captain Higgins, and the mate, Mr. Smith, were typical "Yankees"; Noah the Blacksmith, a Pennsylvania Dutchman. Before the mast were two disgruntled farm-hands, one fugitive from justice, and a Fall River cotton-mill striker. Mr. Frates, the third officer, was a Portuguese; Cooper was a Norwegian; Thompson, boat-steerer, was a St. Helena Englishman; Jim, a Nova Scotian; and August a "Gee" from Lisbon; Smalley, boat-steerer, was a full-blooded Gay Head Indian. All the rest were blacks. Mr. Gomes, the second officer, hailed from the Island of St. Nicholas. Steward was Bermudian. The South Sea Islands, East Indies, Cape Verdes, Azores and Canaries, all were liberally represented in our list. Profane, dissolute, and ignorant they were, yet, on the whole, as courageous and willing a lot as one could desire. Being nearly all islanders, brought up from childhood with oars in their hands, they were eminently suited to the purpose; for boatmen, not seamen, are required in the whale-fishery.

In lieu of wages, a whaler's crew, from captain down to cabin boy, receive each a "lay"; that is to say, a certain proportion of the gross earnings of the voyage. The captain's part may be as much as one seventh, the cabin boy's as low as a two hundred and twenty-fifth — called the long lay.

Now that we were well out at sea, the work on the boats was pushed ahead vigorously; oars, sails, rigging and gear of various sorts were apportioned to each; harpoons, lances, and boat-spades sharpened and mounted on poles. The whale line was stretched and laid in tubs, the kinks being removed by successive left-hand coilings; in some cases the line was tossed overboard and towed astern; for any hitch, when it is racing out of the chocks, fast to a gallied whale, may mean loss of both lives and boat; and both are precious. There were two of these tubs to each boat, containing between them over half a mile of manila rope two and one quarter inches around.

Each boat-steerer conducted the arrangement of his own boat, under the immediate surveillance of his boat-header. In less than a week's time we were ready for whales, and once more the foremast hands were ranged up along the lee-rail midships. This time the boat crews were to be chosen — a far more serious affair than the mere selecting of watches.

The experienced boatmen formed one end of a long line, the green hands the other. Like judges before a dog-bench, the mates strolled up and down the row, now feeling this man's ribs, now making that one bare his arm; occasionally pausing to jerk out a question: "Ever pull in a boat? No? What in hell are you good for? Where are you from? Talk English? Oh, you pulled in Mr. Diaz's boat last voyage, eh? Well, I wouldn't give a Goddam for any man *he* broke in!" The boat-steerers lounged interestedly in the background, now and then proffering suggestions to their heads. When the inspection had been finished, the drawing began. It was evident that the material had been studied critically, for there was but little hesitation and few words were spoken. Now and then there was a grunt when a likely man was lost, or occasionally a mate in a low tone referred a decision to his boat-steerer. When the ceremonial was over, much to their chagrin all but two of the whites before the mast were left for ship-keepers, for they were all very green.

Six men make up a boat-crew. The mate "heads" — that is, commands — the boat, and so is called the boat-header. The harponeer or boat-steerer pulls the forward oar in approaching a whale. After "getting fast" to it, he goes aft and steers the boat, giving place to the mate, who goes forward to wield the lance in the killing. The change is necessary in order to keep the most experienced man at the position of greatest responsibility.

The whaleboat is a "double-ender," some thirty feet long, six feet in beam, with a very pronounced sheer to enable her to ride in the roughest weather. She is sloop-rigged and fitted with a centreboard and a collapsible mast. From the decked-over stern juts a round post called a "loggerhead," around which to snub the whale line. The stem is deeply grooved and set with a roller. Through the "chocks" thus formed the whaleline runs out, being kept from jumping by a slender oaken peg called a "chock-pin." The boat is provided with both rudder and steering-oar, the latter twenty-three feet in length.

Every man before the mast, the boat-steerers and the mates, must do masthead duty. In the Arctic Fishery, a "crow's-nest" is erected to shield the lookout from the severity of the weather, but until a comparatively recent date not even hoops were used in the Sperm Fishery, the "masthead" steadying himself by hanging over the royal-yard; the supposition was that the insecurity of his position would tend to keep him wakeful. The means failed of its purpose not so infrequently as might be supposed, the result usually being fatal. To-day the use of the hoops is universal.

We had arranged our boat-crews one evening after supper, and the next morning for the first time we posted our masthead lookout. Standing on the upper crosstrees with arms dangling over the hoops, great spectacle-like rings bridging the royal-masts breast-high, the green hands passed a miserable two hours; for each graceful dip and gentle roll, which on deck was scarce perceptible, augmented by the hundred feet of sheer mast was exaggerated a hundredfold, and the reeling masts starting on their dizzy downward course appeared about to plunge the very trucks into the yawning depths.

Captain Higgins had offered a bonus of five pounds of tobacco to whoever raised the first whale taken, and with this added incentive four men scrambled up the weather-shrouds, and finding their places in the hoops, with glasses and naked eye scoured the sea eagerly for any sign of whales. On deck quiet reigned, and with all sail set the ship held southward.

After a second night in the steerage, I concluded it was full time to remove to the cabin, particularly as I had noticed that Cooper, now that his supply of liquid friendship had diffused itself, was pondering somewhat on the inconvenience to which he would be put in welcoming another to his already crowded quarters.

The stateroom I now proceeded to share with him had originally been a part of the sail-pen, but when the previous accommodations had proved inadequate it had been annexed to the cabin. It was scarcely larger than a good-sized drygoods box, and an average man could not stand erect in it. Here I had a fore-and-aft berth, the upper one. Our only port opened directly into it, and, worse than its leak, the stench from the bilge reeked up through the trap. It was a simple matter after I turned in for the night to span with one hand the distance from my nose to the deck planking. Hard by my head slept Cooper; beneath me slumbered Steward. In the slight floor space remaining reposed our several sea-chests.

But two men there were on shipboard who did not retire for the night in the same garments they had toiled in during the day. In the morning we all lined up to the water-butt, lashed at the break of the quarter deck, each with towel thrown over his shoulder and tin basin and soap in hand. After bailing the requisite amount of water through the bung-hole by means of a "thief"—a method sufficiently tedious to insure no waste of the precious fluid—each man placed his basin on the edge of the hatch-cover and pursued his ablutions; then took the basin back to its little rack in the ceiling of the afterhouse. For the foremast hands, salt water had to suffice.

There are three messes aboard a whaleship—cabin, steerage, and forecastle. Mr. Smith, Mr. Gomes, and myself sat at the captain's table. At the second table sat Cook, Steward, "Boy," Cooper, and Mr. Frates. The steerage had a table of their own and a boy to attend to their wants. But in the forecastle, each sailor, when his watch was called, reached for his pot and pan, glided aft to the galley along the leeward rail, received from the "Doctor" his chunk of meat or daub of hash, two hot potatoes, and a "tub of slop," and, wiping his sheath-knife on his jeans, sat down on his chest, with the pot of coffee safely propped between his bare feet, and took his nourishment without aid of fork or spoon.

In common with most people, I once had an impression that whaling was a lazy, aimless sort of life, with, of course, occasional spurts of activity, but in the main void of exertion, listless and enervating. I soon had this idea shaken out of me. I have never seen men toil more unremittingly; here at least the hours in which a man might labor were not limited by legislation. Often the work of the day was carried on throughout the night-watches, and Sunday meant nothing at all when whales were alongside or sighted. A storm served merely as pretext for repairing sails not set, and the top-hamper was gone over in the nastiest of weather. New sails were made, spars scraped and slushed, rigging served and tarred, rove and set up. Chafing-gear was braided, the new house over the try-pots remodeled.

Day after day of glorious weather now succeeded, and under a full spread of canvas we bore south and by Bermudas, veered east, and on one long leg made the Western Grounds; passed Azores, passed Canaries, and cruised back again over a part of our course before making southing. Then we struck the southeast trades and bowled merrily along with a following wind, the sun always shining and great banks of fleecy biscuit-shaped clouds hovering constantly over the horizon. Flying fish rose continually at our bows and scattered like autumn leaves to leeward. Dolphins darted in opalescent gleams in our wake; and Cook and I angled for them. On some mornings the sport was exciting. It was Cook who flirted a fish completely over the poop, smashing in a companion-light, the occasion but serving as opportunity for the still further enhancement of my appreciation of the mate's astounding vocabulary.

Every morning before sunrise coffee was served to all hands and the lookout posted. An hour after breakfast, "the Old Man" took his sights and went below and worked up his longitude. Before dinner he again took the sun to fix the latitude. The watches changed at four, eight, and twelve, and at six in the evening. Every two hours a fresh gang relieved the man at the wheel and the masthead lookout. At four, decks were scrubbed down and the pumps sounded. From the time the lookout took their places in the hoops at daybreak till the cry—"Alow from aloft"—brought them tumbling to deck at nightfall, no noise of any sort was permitted. But then in the second dog-watch the hands gathered about the windlass or stretched at full length on the forehatch; smoked, yarned, and indulged in horse-play as they desired. The green hands perhaps studied the points of a dummy compass, or under competent teachers fingered the ropes, committing them to memory. Two misfit concertinas every night sent up their dismal wail to a tune which never varied, often keeping time with the strokes of a couple who pounded up hard bread in a canvas bag to be mixed into a molasses mush for their watch. The boat-steerers lounged on the workbench aft the try-pots, whetting harpoons and swapping stories.

Mr. Smith and I often would sit on Cooper's chest before the afterhouse, tying knots, talking of home and of past gastronomic achievements. We had drifted into the habit of meeting two or three times a day whenever there was opportunity. Early in the cruise, when he was still convalescing from beri-beri contracted on the previous voyage, we would sit together out on the main upper topsail yard through the quiet hours of the afternoon. Surrounded by great banks of billowing canvas and shut completely off from all view of the deck, Mr. Smith would expatiate at length upon the habits of whales and his own observations of them.

"There are only two kinds of whale," said Mr. Smith. "One of 'em is the Sperm Whale; the rest of 'em is the other. The Sperm Whale is mainly valuable for his oil (sperm-oil, you understand); only has teeth on his underjaw, like a cow, fights at both ends, has one forward spout, and lives only in warm country. Now Right Whale oil ain't worth beans; you hunt him for bone; 's got a whole sieve made out of slabs of bone in his mouth 'stead of teeth. Then he only fights with his flukes; but you bet he can use them pretty lively. Never known of a Right Whale's crossing the Line. Swallow Jonah? Humph! Well, a Sperm Whale could 'a' done it, but how'd you like to swaller a woolly worm? No wonder it went agin' his stomach."

Again we frequently met in the larboard boat when decks were flushed at night. Usually he would be chewing a piece of raw salt codfish gleaned from a box Steward had stowed under one of the spare boats. It was from this point of vantage I was startled one day by an unusual commotion forward. We had driven into a school of porpoises, breaching and gamboling all about us. There was a scramble for harpoons, and in a moment the boat-steerers were out on the back ropes in a race to get fast. There were two misses before Smalley drove his iron into one, for a porpoise's movements are erratic as a rabbit's. Flopping and bleeding like a stuck pig, the animal was hauled to deck. That night we had the first fresh meat in weeks, and for two days we feasted on porpoise steaks and liver. The steaks are a trifle oily, perhaps, but the liver is as good as any. Sea-pig he is called by the sailors. The cook tried-out the head and jaw oil, the cabin-boy saved the teeth, and Mr. Gomes cut the crotch from the flukes for a talisman and nailed it to the cutting-stage—the first catch of the season.

No labor except working ship and keeping lookout is required on Sunday unless a whale be sighted or there is oil to care for. The ship presents on that day long lines of fluttering garments strung from davit to davit; patched, quilted, abbreviated, till all semblance to the original scheme is lost. Laundry work attended to, the next thing in order seems to be the cutting of hair. The less energetic sleep; others read the magazines furnished by the several charitable societies ashore. Some write letters, some patch and mend, but always aloft four pairs of eyes scour the sea for whales, and the ship holds her course.

One Sunday morning a tremulous "Bl-o-o-w!" from aloft brought me to my feet, trembling with excitement. "There she blo-o-ws!" thundered from all four hoops. "Ah blo-o-w!" Captain shot from the companionway and bolted up the main shrouds. "Where away?" shouted Mr. Smith, dropping his piece of salt codfish. "Two points off the lee bow, sir; school Sperm Whale," answered Thompson. "Keep her steady, Mr. Smith," called Captain, squirming by the maintop. "Ah bl-o-ows!" chorused the mastheads. "There! She white-waters!" "What time is it?" shouted Skipper from the topmast shrouds. "Quarter past nine!" yelled Cook, craning by the shoulder of the helmsman.

Feverishly the decks were cleared, the men gathered in their lines of washing, the boat-steerers gave the last few touches to their boats and cleared the line-tubs. The whales now had sounded. In suspense we leaned over the bulwarks and gazed to windward in the direction they had gone. At last, after what seemed an interminable wait, an instantly stifled "B-l-ows!" sounded from aloft. "Lower away!" came the order. In the midst of a breathless excitement, three boats were dropped, the sails hoisted, and a long beat to windward began. The whales spouted but a few times and then went down. When they broke water for a second time, it was at least eight miles from ship, and a signal was set to recall the boats.

In the *Sunbeam's* log-book is the brief but significant entry, "Raised whales this a.m. going quick, lowered, no success, come on board!"

We shortened sail at sunset for the first time, and afterwards we cruised through the day, and at night carried barely enough canvas to keep our steerageway.

HARPOONING A PORPOISE
In the collection of the New Bedford Public Library

CUTTING-IN
An 85-barrel whale alongside the Sunbeam, 1904
In the collection of the New Bedford Public Library

BAILING THE CASE
In the collection of the New Bedford Public Library

FAST BOAT

When two or more boats fasten to the same whale, the ranking officer's boat leads away and the others tow astern
In the collection of the Mariners' Museum, Newport News

CHAPTER II
THE BLUBBER-HUNTERS, *concluded*

WE raised our next whale at sun-up on another Sunday some weeks later and almost before the first long-drawn "Bl-o-o-w" from aloft had ceased its echo our boats had dropped like shadows on the surface of the desert ocean. One moment the decks had been hushed and quiet. The next it was as if a squall had struck us—a hurricane of orders preceded a wild stampede, and underfoot was instantly strewn with a tangle of braces, sheets, and halyards thrown by the mates from the pins. The men reached and hauled, the mates slacked away, the great yards swung, and the old bark slowly came about. Shoes were kicked from feet, and while we still luffed, the bow boat took to the water with scarcely a ripple, and the boat-crew swarmed down the falls and dropped catlike into their places.

With a whir of tackle the waist and starboard boats followed; so closely they seemed almost to strike the water in unison. Carried away in the excitement of the moment, I forgot a pair of burned hands I was nursing, and brushing aside one of the men, I slid down the davit tackle, and found myself in the mate's boat, struggling with the rest of the crew in the maze of gear.

Wallowing precariously in the heavy wash, we balanced, on gunwale and thwart, and stepped the awkward swinging mast, then made fast the stays. The sail was hoisted and peaked, the sheet paid out, and with the boom sousing through the water, we started down the wind full thirty yards ahead of Mr. Frates. The boats were soon widely scattered. Occasionally another could be seen when the seas at the same instant tossed both of us high. The *Sunbeam* in the distance showed only her top-hamper. Of the whereabouts of the whale we were advised from time to time by her signals aloft. The colors soon had disappeared from the main-truck, indicating that the creature had sounded. So we held our course unaltered till the weather-clew of the foret'gallants'l told us he had broken water far off to windward, and there was nothing left but to in sail and pull for it.

For four hours or more we tugged at the oars, changing our course from time to time to suit the varying whim of the whale; several times we were almost within "darting distance," when he turned flukes and sounded. So we pulled down, perceptibly gaining nearer, and the chase every moment becoming more and more intense. We strained at the oars till the mate, crouching in the stern-sheets, was blurred to our sight, and brilliant specks of light danced before us; our throats were parched, our hands were bleeding.

Suddenly, "Stand by your iron, Tony!" and from somewhere back of us came a faint, sonorous whistle. I twisted my neck, and caught a glimpse of a dark mass cleaving the water a hundred fathoms ahead. Our oars bent like reeds, the boat leaped ahead like an animal under the lash. Stooping low over the steering-oar, Mr. Smith jerked out crisp words of encouragement to us. Again the whale spouted nearer, and then, for a third time, that long-drawn whistling exhaust; and the humid vapor escaping from the pent-up lungs drifted like a mist over the boat, and we felt on our necks its dampness. The rankness of it was still in our nostrils, when there was a swirl and a rush, and the huge monster breached clear of the sea, and with mighty flukes tossed high, for an instant was silhouetted against the yellow of the afternoon sky; then with a deafening roar of displaced waters, disappeared beneath the surface, just as another whale coming from we knew not where, broke water under our very bows. "Give it to him!" yelled the mate. Tony gave a frantic lunge, and the harpoon was buried to its hitches; overboard went the second iron.

"Fast, by God! Peak your oars! Get out of the way of that line! Empty some water on that tub, you damned lubbers!"—and with a dash over the tottering thwarts, Tony and Mr. Smith shifted places. A jerk pitched those unprepared in a heap to the bottom. For an instant our stem was sucked under, and we shipped a barrel of water. Then we were off with the speed of an express train, with the water pouring in a sheet over our bows and sifting the full length of the boat, till the last of us was drenched to the skin. Writhing and squirming from tub to loggerhead, from loggerhead to chocks, under the kicking-strap and along the channel formed by our peaked oars, the line whistled and tore, sometimes with a rumble like the roll of drums, sometimes with the wailing shriek of a siren—whistled till it smoked, yet still the boat was being lashed through the water like a fly on a trout-line, and behind us rolled up a wash like that of a steamboat. The water boiled at our bows and eddied by our quarter, and we left a line of suds behind us far back to the horizon. Buffeted from side to side by the oncoming waves, the boat creaking and quivering under the impacts, we tore in an arrow-like flight due westward, till the ship disappeared in the distance.

Tony stood by the stern-sheets, a canvas "nipper" in hand; from time to time he threw a turn of the line over the loggerhead, sometimes cast one off, but only to go on again the instant after. And so we held our speed without let-up; for though, after a while, the whale slackened his first mad pace, the line also went out slower. And we, crouching in the bottom to steady her, bailed constantly to keep from filling with the influx over the bows. Slower and slower the line surged, but the stern tub was emptied and the waist tub was being drawn on heavily. Then came a moment when it ceased running through the chocks, and the boat began to lag.

"Stand by to haul in line!" came the order, and getting out of our cramped positions, we grasped the now inanimate rope and hauled and strained with feet braced on the thwarts, winning it back inch by inch, painfully and slowly. Tony held what we made with his turn on the loggerhead. Span by span, fathom by fathom, the line crawled in, and we coiled it in the stern-sheets.

Gradually we hauled up to the whale, and our nerves tingled again in suspense. With knee braced in the clumsy cleat, Mr. Smith stood ready, his lance poised high. We dragged up abaft the churning flukes, then got out our oars and pulled till we scraped the barnacled flanks, "wood to blackskin." Almost bending double, Mr. Smith drove in the lance till it brought up at the socket, a full six feet of cold steel; then for an instant he "churned" his weapon. The great fanlike flukes lifted from the water; gently they seemed to tap the surface, but the report was like that of a cannon, and the other two boats, two miles away, caught its reverberations. "Vast pulling! Stern all! Peak your oars!" And we watched all that coil of line, fruit of an hour's toil, go over the bows again, all to be hauled in anew.

But our quarry was now spouting blood and visibly growing weaker. Four wearisome times, hand over hand, we hauled up to and lanced him. Each time he carried out less of our line. He no longer swam with the decision of his first frenzied flight, but frequently altered his course, spouting oftener and thicker and with visible effort. Behind him he left an ever-darkening trail of crimson. Then, when we were still at some distance, as though goaded afresh, he churned his flukes, and, with vast form listing to one side, tore with some suggestion of his original pace in a large arc around us. Suddenly he veered sharply; then, with a horrid inward convulsion, a stream of clotted crimson gushed from his spiracle, and the great carcass turned fin up, with the seas lapping over it—lay just awash, a huge, shadowy, undulating mass, with no more semblance to a living creature than had the seaweed drifting by it.

Already the scavengers of the deep were gathering, their sharp fins cutting the water knife-like all about us. Not a moment could we halt; the day was all too short for the task before us. Reeving a short warp through a hole cut into his spout, we passed a line from one boat to another, and, all a tandem, began the long hard tow to ship.

Luckily, the *Sunbeam* crowded sail, and with a favoring wind bore down to us. After what seemed an interminable pull, we dragged the carcass alongside and passed up the tow-rope. With a rattle and jangle the fluke-chain was belted, and heaving away at the windlass, the whale was soon fast alongside.

As we rounded the *Sunbeam's* stern to get under our davits, the grateful aroma of coffee greeted our nostrils, and with renewed energy we clambered to deck and triced the boats up to their places, swung the cranes, and lashed the gripes.

"Dinner, all hands! Dinner, Cook, dinner!" There was a stampede aft, and with a clattering of pots and pans the half-starved crew mobbed the galley and received double rations.

A whaleship "cuts-in" a whale always over the starboard side. To admit of this, three of the four whaleboats are suspended from the larboard side, while to starboard, over the gangway, is lashed a long platform, or cutting-stage. This is lowered from the ship's side and boomed out some ten feet over the dead whale. From this platform, which has a handrail, the mates work, cutting at the blubber with long-handled, keen-edged spades, similar to those used to clear ice from sidewalks.

If the weather is not too rough, a ship will "cut to windward" (with the whale on the starboard side toward the wind). By so doing the wind pressure on the sails will counterbalance to a greater or lesser degree the weight at the cutting-falls, and so tend to keep the ship on an even keel. The strain at the mainmast is terrific. It is said that on occasion the weight of a "case" has been known to cause the foot of a mast to crush through a vessel's backbone, so scuttling her.

The cutting-tackle consists of a cluster of gigantic blocks made fast to the main-top, through which are rove the two falls, each suspending a heavy block and blubber-hook. This cluster of blocks is braced well forward with a jigger, in order that the two hooks may swing directly over the blubber-room at the main-hatch.

The blubber, except for a filmlike coating called "blackskin," easily scraped off with the thumbnail, is the only outer covering of the whale. It is of a fatty nature, but is of very close texture and exceedingly tough. This separates readily from the flesh beneath, so that, in general, only vertical incisions are made with the spades along a line termed the "scarf," and the lift of the windlass rips the blubber from the carcass much as the peel is skinned from an orange, requiring only an occasional jab from the spade to keep it free. The whale rolls over and over in the water as it is unwound. The mates on the stage hack with their spades a corkscrew line about the rolling body, and the heave of the windlass tears away the blubber. When the end of the strip has been hauled chock to the maintop, the third mate at his station in the waist, with a long boarding-knife punctures two holes through it close to the deck and at some distance apart. Through these a chain strap is rove and the second tackle hooked to it. Then, all hands standing back from the gangway, with a few well-directed lunges the mate severs the mass above the newly fastened tackle, and to the lusty shout of "Board ho!" the great weight of the first "blanket piece" swings inboard, sweeping any luckless obstacle from its path, and dragging about with it those who attempt to steady it down the hatchway.

The day was too far advanced when we got our first whale alongside to do more than get the fluke-chain in place and prepare for an early start in the morning. During the night the wind freshened, and by morning was blowing half a gale. The ship labored badly, with the whale's enormous bulk jerking and bumping alongside. After the expenditure of an infinite amount of patience and labor, the first blubber-hook was embedded at the base of the left fin and a semi-circular cut made below it. With the carcass yawing and twisting, the seas continually breaking over it till it was often submerged completely, not one thrust in six of the cutting-spades was effective. To get the hook in place, it was necessary to send down a man.

"Who's overboard?" Mr. Gomes asked.

From a dozen volunteers, one was selected — a big, hulking negro, distinguished from the rest by a wide cotton rag he wore under his jaws to keep his hat in place. Giving the man chance to remove only his coat, a "monkey rope" was fastened about his waist, and he was dropped sprawling on the slippery, heaving flank of the whale. About him hovered innumerable sharks, gliding like shadows alongside, now and then tearing away a hunk of flesh and flopping silently off, pursued by others equally ravenous.

With lines a couple eased the weight of the enormous blubber-hook, while the negro groped for the hole made near the flipper. The mates, with backs to the hand-rail, fought off the sharks from too close proximity. The seas constantly broke over him, and twice the man was washed from his task, and was only preserved by the rope about his waist from being ground between the whale and the ship's side. Then, with the hook in position, the windlass crawled up on the slackened fall so slowly that the poor fellow seemed to be under water for nearly a minute, hugging the hook to its place. Suddenly the ship yawed, and then with a slap and jerk the tackle tightened. There was a hollow rending sound, and the blood-dripping semi-circle of blubber lifted from the carcass, with the disjointed fin flopping at its end. As the ship rolled far over, the carcass raised a third of its bulk from the water, then settled suddenly back with a sullen splash, and a straightened blubber-hook jerked high in the air and fell with a crash on deck. The whole wearisome process would have to be repeated.

A second hook having been fetched, it seemed about to hold, when a sea a little larger than common broke completely over the stage, and, lifting the whale bodily, crashed it up under the stage-planks, and would have carried away the mates but for their lashings; as it subsided, the whole weight of the whale was thrown suddenly on the tackle, and the second blubber-hook was left dangling in mid-air, so much junk.

The officers by this time were all well out of patience, and the crew naturally had to bear the brunt. Orders followed each other in a torrent, and were no sooner given than countermanded, the whole interspersed plentifully with curses for the fancied stupidity of the crew. A little late in the day, perhaps, it was decided to tack ship and cut to leeward. Our unwieldy burden, acting as a "drug," made it impossible to get up headway sufficient to put over directly. So over two hours were spent wearing ship. For a long while it looked as though even this was not feasible, but at last we got about, and were well rewarded by the comparative quiet in which our "cut" lay. We swayed the stage a trifle higher to allow for the starboard list of the ship, and the wind having abated considerably since early morning, but for the heavy ground-swell no easier cutting could have been desired. The only remaining blubber-hook—a new one and much larger than the others—was broken out, and for the third time a man went overboard to secure the hook in place.

It would seem that we had received our due of accidents. Yet no sooner had the blanket been well started toward the maintop than a sudden lurch gave an extra strain to the fall, which was an old one. It parted and the block, tackle, and our third and last blubber-hook went clanking overboard.

While Cooper hewed an old-fashioned toggle from an oaken plank, the rest of us sat down to a glum meal in the cabin. Skipper even omitted his customary joke— a threat to tie his napkin around the cabin-boy's neck, when that individual had forgotten to fetch it. We faced at this juncture the dismaying prospect of cutting-in a whole voyage without a hook, and the outlook was anything but cheerful. So dinner was bolted in a gravelike gloom, and the work on deck resumed directly. A chain strap having been with considerable difficulty shackled about the lower jaw, the body was rolled over, and there, pinned by the strap in the angle of the jaws, were the missing block and hook. How they had remained in such an insecure position is a mystery. But no time was given to speculation. A fourth man plunged recklessly overboard without pausing to make fast the monkey-rope, and three more sharks gave up their ghosts.

Seven hours' arduous labor had been performed and our job was not yet started! But if all things before had seemed to hinder, all now seemed to facilitate matters. A new cutting-fall was rove, the hook was lowered from the cutting-stage and it dropped directly into the hole without necessitating a man's going overboard. The blubber lifted easily; the ship rode quietly; the work progressed without a hitch. The carcass was decapitated and the head secured astern. The body rolled over and over, weltering in gore, the great dripping blankets followed each other up the alternating falls.

Mr. Frates by the gangway sliced with his boarding-knife and swung the pieces inboard. Cooper perspired over the squeaking grindstone, muttering curses for those whose continual demand "Sharp spade! Sharp spade!" kept him "humping." Between decks, in the blubber-room, one watch stowed down the "blankets" and cut them into smaller "horse-pieces." Above, the other watch steadied the huge chunks by the hatch-coamings, and lugged the heavy block and chain back to the gangway, slipping and sliding in the slush and gurry.

Moving too deliberately from the path of one of the inswinging blankets, one of the green hands was swept from his feet like a tenpin, carried across the deck and slapped down the hatchway. For the rest of the season he was on the sick-list.

About us in a great circle the waters were crimson with the outpour of blood from the carcass. The sea fairly boiled with monstrous sharks battling among themselves for the detached fragments. The work was progressing smoothly, when the ship suddenly lost her headway, and a faint, all but imperceptible, tremor was felt in the deck. We had been barely under steerageway; now we made sternway. It seemed almost as if some curious under-the-surface current had seized us. The whale swung from under the stage, making further work impossible. The men exchanged mystified glances; the mates stared aloft, but everything seemed drawing properly. A shriek from aft broke the silence, and Steward struggled from the rear of the galley, dragging hapless Cook by the ear. "This damned nigger been an' harpooned a shark!" Sure enough, furiously lashing the water into a foam astern, a thirty-foot shark was just succeeding in freeing himself from the cook's investigating iron, made fast to a cleat at the taffrail. The ship, freed of the shark, serenely gathered headway, the whale swung in, and all hands resumed their labors.

The spiral cutting progressed to a point midway between the hump and flukes; then, after the body had been disembowelled and searched for ambergris, two vertebræ were disjointed and the carcass cast adrift. Hauling the remnant partially from the water, the flukes were severed at the small, and, freed of the chain, followed the denuded carcass down to the sharks. To the exultant cry of all hands, "Five and forty mo-o-ah!" the shank of the tail smashed over the sheer-plank and spun across the deck.

We hoisted the junk, the lower half of the forehead, and made it fast to the lash-rail, aft the gangway. Then came the case, the real lift of the day. Its twenty tons or more brought the starboard scuppers down to the water-level; the ship's hull creaked and groaned under the strain. Half on deck, half by the board, it was secured fast and the stage hoisted out of the way.

"All hands aft to splice the mainbrace!" came the welcome call. And further work on the whale was stayed till observance was paid to this time-honored custom. It may be that the hearty bawl when the last of the cut swings inboard is as much due to the anticipated grog as to the five-and-forty barrels nearer "full ship." Standing by the quarter-deck, our skipper dealt out the stimulant from a cracked and brimming pitcher. Filing past, each man in turn received the tumblerful, and tossing the raw stuff at a single gulp, passed forward with watery eyes.

"Bailing the case" is perhaps the most interesting of the several processes peculiar to whaling. After "splicing the mainbrace" some portion of the litter had been cleared away from deck and the cutting-falls sent down. A hurried supper was eaten, after which one watch was sent below and the other turned to preparing the try-works, cutting horse-pieces, and lastly bailing the case. The waist was lighted with a few thick-globed lanterns, which diffused a feeble radiance over the scene. Forward a cresset of burning scrap flared above the try-works.

The tail of a beaver, the hump of a camel, and the case of a sperm-whale have each the same function — the hoarding-up of reserve nourishment against a time of fast. Fatty and unctuous, glistening and pearly white, the cavernous reservoir lay opened before us like some vast comb of honey, trickling its stored-up treasure over the sullied planking, turning it to purest snow. Stark naked, three negroes climbed into its tank-like interior, and wallowing to their waists, with knives and scoops, half cut, half ladled the barrels of pulpy, dripping substance from its cells. With tubs, buckets, and pails, an improvised bucket-brigade passed the prized contents forward to the try-pots, where two bronze-like figures, standing in the capacious kettles, with groping fingers tore the oozing pulp to shreds.

Delving deeper and deeper with an eagerness requiring no encouragement, the bailers labored without cessation. The try-pots were filled, but still the supply held, till thirty barrels and more of pure spermaceti stood in brimming tubs along the bulwarks. The scuppers had been stoppered, and the deck was awash with gurry and congealing case-matter. Through this the men splashed and slipped, and with "save-all" and shovel scooped the precious leakage and poured it into tubs.

Under the try-pot fires were started, and the flames leaped hungrily high above the funnels, throwing a lurid glare over the shifting scene. Above, the wan ghostly sails flapped and glowed, the flames contorting wildly in the back draught caused by the flapping. Black toiling figures teemed like ants about the decks; and all made a picture the weirdness of which suggested a transcript from the nether world. Like a presiding evil spirit, Smalley's dark face shone in the intense heat before the works, as he forked the minced "books" of blubber and soused them in the seething cauldrons.

Watch relieved watch, but all through the night the work went on. The horse-pieces were minced, the tried-out "scrap" was fed to the fire. Black smoke belched from the stacks, darkening with a thick soot the rigging aloft and the near-by bow boat. The tried-out oil was bailed to the deck cooler. More blubber was fed. The men, passing by, helped themselves to choice bits of well-fried scrap. A pungent, sickening odor of burning blubber burdened the air.

In the morning we sighted another New Bedford whaler, also boiling, the schooner *Eleanor B. Conwell*. But before there was opportunity to speak her, we again raised whales, and by noon had another alongside. In a drizzling rain we finished cutting-in and stowed the damp blubber between decks.

The second morning the skipper of the *Conwell* came over with a boat's crew and "gammed" with us; during his stay on board, his crew joined in the labors with ours, and as they worked they discussed the comparative merits of the two ships, their captains, and the fare.

Mr. Gomes, myself, Tony, and a boat-crew lowered in the starboard boat and cruised till nightfall with the schooner. In the heavy ground-swell which was running she made bad weather of it, the big boom thrashing from side to side, jarring and racking the whole craft. The steward was the only white man aboard, and he made a melancholy tale of his trials with the captain, who, he said, permitted him no molasses to cook with, no yeast for his bread, and as for butter, "Why, the damned Gee eats lard on his bread, and thinks a white man oughter."

The crew were a rather wild lot, and already underfed; their mutiny a few months later might then have been predicted. I came back to the *Sunbeam* minus a jack-knife and with my pockets full of gingerbread, the cook having helped himself to molasses in the captain's absence.

Our pots could try-out about two barrels of oil an hour, and at this rate we now had perhaps fifty barrels of almost boiling oil in the large metal coolers between decks. Driven aft by the heat, the cockroaches that night literally swarmed the cabin. Time and again we were awakened by their running across our faces. Pulling on my boot hurriedly in the morning, I encountered no less than six of them.

Almost under the try-pots, and with the floor buried in a heap of musty and mildewed garments, wet and oily from constant duty overboard and contact with the dripping blubber, the forecastle was a veritable hell.

The wet blubber between decks began to rot within twenty-four hours after being stowed down. Which was the more obnoxious, the burning scrap on deck, or the decaying blubber below, is difficult to determine. That night I attempted to sleep on the roof of the afterhouse, propped against a line-tub. At 2 A.M. there was a miniature cloudburst, and I went below and spent the remainder of the night in a temperature of well over 100°.

If anything, it was hotter below the second night. I had planned to sleep under one of the spare boats, but I found Steward and Cook occupying the only suitable places. So I determined to risk another night on the galley. There was scarcely any breeze blowing, but the flapping of the spanker to the ship's pronounced roll made an intermittent draught, which fanned my face, keeping me awake. Half a mile off our beam the *Conwell* pitched and rolled, boiling her oil. Her sails were ruddy in the light from the try-works, and cast a fitful reflection over the water. I went to sleep with its glow dancing confusedly before my eyes, and, perhaps as a consequence, was tortured by the most blood-curdling dreams. In the end a realistic blow on the head brought me to my senses, and I wildly grabbed the davit I had struck against, barely in time to save myself from going overboard. We had tacked ship earlier in the night, the first time we had altered course in twenty-eight days, and the freshening breeze having tilted the deck to a steeper angle I had rolled in my sleep completely across the house. Gathering up my blanket, I started below, noticing with a shudder as I passed the companion that the man at the wheel was fast asleep. There was not another person aft.

We raised whales again within the week, and again the five pounds tobacco reward fell to Thompson. The commotion of the other occasions was repeated. Amid creaking of blocks, flapping of sails, shuffling of feet, a Babel of orders and the crew's hearty "Aye, aye, sir!" covers were removed from line-tubs, casks and stores hurriedly lowered to the hold, breakfast swallowed, and in fifteen minutes the boats were sent down and a long pull to windward ensued.

On account of my hands, which had been badly barked at the previous lowering, I decided this time to remain aboard ship. After the boats had lowered I slung a pair of field-glasses about my neck and joined Captain in the hoops. The day was hot and sultry, and the water reflected the sun's blinding glare into our eyes. Way below us the pygmy seamen scurried about deck, working ship, and setting the various signals to keep the boats advised of the whales' whereabouts. Evidently we were pursuing a large school this time, for the spouting was continuous. So the colors stayed hauled to the main-truck. After a long chase, one of the boats got fast, and the panicky flight of the whale soon carried it all but beyond our vision. I went to deck and stretched my legs and again took my post at the lookout in time to witness the flurry. While we were absorbed in the culmination of the distant tragedy, Cooper announced from deck the sighting of another whale. Sure enough, not two hundred fathoms off our quarter, another leviathan was steaming up, logging about two feet to our one, and heading directly across our bows. He swam so near, not varying his course, that a collision seemed imminent, and recalling the fate of other vessels that had been rammed by whales, Captain Higgins became alarmed and ordered the hands to make all noise possible. So they pounded the deck and the water-butts, pumped the squeaking windlass, clanged the ship's bells, and banged tin pans. Amid the clatter, but with a dignity consistent with his proportions, the whale settled from sight and passed under our bilge and away. Under a fair wind we now ran down and picked up the waiting boats, and so got the second "cut" alongside.

One day, shortly before the end of my voyage, we raised another solitary bull, and took to the boats on sighting him. Our boat had lowered first, and we were away before the next boat, Mr. Gomes's, swung from the cranes. Just before he took the water the forward tackle fouled, and the boat's stern swung under the *Sunbeam's* counter, just settling in a sea. The stern-post and steering-oar brace were smashed to splinters and the rudder put out of commission. We left Mr. Gomes stretched at full length over the cuddy-board, furiously rigging a jury steering-gear, and jabbering, while he worked, a stream of Portuguese oaths, impartially directed at both his crew and the job. But the chase was a long one, and when we finally got an iron to the whale it was from Mr. Gomes's boat, and later Mr. Frates closed in and fastened.

It was toward the end of the afternoon that we found ourselves racing abeam of Mr. Gomes and within earshot of the labored spouting of the whale.

Mr. Smith and Smalley quickly shifted places (we had traded boat-steerers since Tony had missed a big whale). The boat pulled up beside the fugitive. Quickly the varying orders came: "Give way all! — 'Vast pulling Three! Pull Three! — Stern Two! — 'Vast pulling Two! — Give way all! — Hit her up lively there! — All together now! — Steady! Ste-a-dy!" Then the great mossy hump was swimming so close to us I could have reached over and touched it; and bobbing some fifty fathoms astern, with drawn, anxious faces, trailed the other two boats, mere puppets in the drama. Suddenly we heard the sharp suck of the lance, then a hoarse, "'Vast pulling! Stern all! Stern all!" We obeyed just in time! Under us the great flukes lifted with a crash, and we canted off and nearly foundered. I had barely got my oar back in place when the whale broke water again, and with an exhaust like the bellow of a bull, cut across our bows. Instantly the drawn lines trailing behind him to the other boats slipped over our blades with a deadly grip, and began to creep up the looms of our oars. The boat listed and the water rose to the gunwale. We struggled vainly to liberate ourselves. Amos tried to swing her off, but we were being carried broadside with such force that the steering-oar was of no avail. "Cut the line!" yelled Mr. Smith. Kite, the man at my back, dropped his oar, and drew a sheath-knife. The line was whizzing over the oars with a rumble that sounded exactly like a wagon going over a plank bridge. We clung to them for dear life. Kite raised the knife and slashed. Not having allowed for the velocity at which the line was running, his edge turned, and the knife buried itself an inch in his thigh.

All this while the fast boats were tearing down on us, yet it seemed as if they would never realize our predicament and slacken line.

I saw Kite with his knife poised for a second jab. But he had let go his oar in the meantime and the whale-line, pulling over it, brought the handle with a resounding blow against the side of my head, and I lost interest in all immediately subsequent proceedings. I came to a few minutes later to find Alfred, the after-oar, emptying a bucket of sea-water over me, and the boat floating tranquilly on an even keel. The other boats were clustered about in a "gam," and the lifeless whale was drifting quietly up to windward.

That was the end of my story. We had very good whaling, so good that it was necessary to pump overboard several hundred barrels of fresh water to make room for our oil. Soon after the events recorded, we put into Brava to recruit, and there I left the *Sunbeam*. Twenty-three months after she sailed from New Bedford she made port again, a full ship.

CHAPTER III
THE GREENLAND FISHERY

ACCORDING to early accounts, the people of a number of different nations had been successful in capturing whales in olden times. But the first systematic commercial fishery was established by the Basques and Gascoynes, who whaled so successfully in the Bay of Biscay during the twelfth, thirteenth, and fourteenth centuries that the whales were almost driven from their shores. As a result, their fishery fell off in importance during the fifteenth century, although by this time they had extended their cruises to the shores of Iceland. Early in the sixteenth century they established the Newfoundland Fishery.

When the Dutch began to fish Spitsbergen, after its discovery by Barentz in 1586, Basque harponeers were hired, and made up a considerable part of every crew. After Thomas Edge's initial voyage in 1611, the English pursued this same fishery, and they also engaged Basques to act as harponeers. The first English whaling voyage on record had been to Iceland in 1598 (Scammon, p. 189), when several ships were fitted out by the merchants of Hull. Spitsbergen was named Greenland by the British, as they then believed that land was continuous to the westward, connecting with the eastern coast of what was then called "Old Greenland," now Greenland. So the British term "Greenland Fishery" applied to all the seas lying to the eastward of Greenland, and included Iceland and Spitsbergen. The more recent fisheries to the west of Greenland were named either "Davis Straits Fishery" or "Baffin's Bay Fishery." As Spitsbergen diminished in importance, it came to be known as "East Greenland." The Dutch from the start distinguished between the two, calling one Spitsbergen, the other Groenland.

The Dutch, the English, and in lesser degree the Danes, the Germans, and the French, proceeded to prosecute the Spitsbergen fishery. The whales along shore were at first so abundant that the ships remained in the harbor and boats were lowered at the anchorage, frequently capturing all their whales within sight of the ships. At first there was considerable jealousy and altercation among vessels of the different nations gathered on the grounds. But finding that this was not conducive to successful whaling, they at last allotted among themselves the various harbors and bays, and continued to fish in comparative harmony. In all these fisheries, the Basques were the harponeers and boat-headers.

The Dutch, by their greater zeal, soon outstripped the other nations. They founded a town called Smeerenberg in one of the bays, and erected try-houses, store-houses, and living quarters for the shore crews. During the winter months this town was deserted, for the fishery was north of the Arctic Circle within eleven degrees of the North Pole. Only twice in history did men successfully winter there. But they made up for their winter absence by double-manning their ships, and fishing twenty-four hours a day in the constant daylight of the Arctic summer. At one time, shortly after 1637, the Dutch are said to have had upward of three hundred ships and eighteen thousand men at Spitsbergen. The war between England and Holland put a stop to whaling in 1653, but in the early 1660's the Dutch had an even greater fleet of over four hundred ships, while England's had dropped to an average of less than one in a season.

In 1719 the Dutch made their first voyages to Davis Straits on account of the growing scarcity of whales in Greenland waters. There had been no English whaleship at sea for a number of years, but in 1725 the South Sea Company, originally founded in 1711, was revived, and built twelve new ships of about three hundred tons each, which were sent to Davis Straits. Having neglected the industry so long, and having no experienced whalemen of their own at this time, they took boat-headers and boat-steerers from the Friesian Islands, just as a century before they had hired Basques, and just as fifty years later they employed Nantucketers. These twelve ships took twenty-five whales the first season, and, encouraged thereby, the fleet was increased until in 1730 it consisted of twenty-two vessels.

Speaking of this period in the English whale fishery, and of the indifferent success of the fleet which was kept at sea only by substantial bounties, Scoresby says: "While the subjects of Great Britain performed a voyage so distant and practiced arduous operations in the polar regions, the colonists in America had the advantage of conducting the fishery more immediately at home. Hence we find many notices of their successful labors in this speculation."

THE GREENLAND FISHERY

Nantucket had one whaler in Davis Straits in 1732, in which year, the English South Sea Company having failed, the English sent out but two ships. (Scoresby, p. 72.) In 1737 there were between fifty and sixty American whalers in Davis Straits alone. ("Fishery Industries of the United States," p. 94.)

Finding that a bounty of twenty shillings a ton was insufficient encouragement for her whalemen, England increased the amount in 1740 to thirty shillings, as she was anxious to preserve her whale fisheries as a school for a naval reserve. Again in 1749 it was necessary to increase the bounties, which by this time totalled forty shillings per ton. Under this stimulus the fishery began to gain, until "the combined fleets of England and Scotland in the year 1752 amounted to forty sail." (Scoresby, p. 75.) Previous to this period Scotland had not engaged in the fishery, but soon after this Dundee, Greenock, and Aberdeen, began to take an important part.

It should be borne in mind that during this period England was actively at war with France most of the time, and that the whalers were particularly open to attack. This was sufficient to discourage any but the boldest from investing in so hazardous an enterprise.

However, it was not until the Revolutionary War drove the American Whaler temporarily off the seas that the English whale fishery regained the importance it had held in the early seventeenth century.

In 1782 England's Greenland fleet consisted of thirty-eight ships; in 1784, eighty-nine; in 1785 there were one hundred and forty; and in 1790 over two hundred. To understand this sudden increase, reflecting a more successful pursuit of the fishery, it is only necessary to know that practically every whaleship in the Nantucket fleet — one hundred and thirty-four out of one hundred and fifty — was captured by the British early in the Revolutionary War, and that every whaleman captured was given the privilege of deciding whether he would continue whaling under the British flag or go to prison. Since they were not required to bear arms against their country, most of the Nantucketers chose to serve.

John Adams, writing from France in 1778, urging protection for the whaleman and retaliation upon the English for their depredations, says in part (A. Howard Clark, "Fishery Industries of the United States," p. 123): "The English, the last year and the year before, carried on this Fishery to very great advantage, off the River Plate in South America, in the Latitude Thirty-five South and from thence to Forty, just on the edge of soundings, off and on, about the Longitude Sixty-five from London. They had seventeen vessels in this Fishery, which all sailed from London in the Months of September and October. All the officers and Men are Americans."

To quote a contemporary British authority (David Macpherson, "Annals of Commerce," vol. 3, p. 590, London, 1805): "Another branch of trade also took its rise in some degree from the war — the Whale Fishery [American] being given up in consequence of the war, many of the harpooners were induced to enter into the service of British merchants who fitted out vessels for the Newfoundland and Southern Whale Fisheries. For the latter [sperm-whaling] which was quite a new business for this country, there were equipped fifteen vessels of about 170 tuns, each carrying four American Harpooners." So started the English Sperm Whale Fishery. According to another English authority (Beale, p.143), the first English vessels to attempt Sperm whaling started to the South Seas in 1775. Ten of these ships appear to have been captured American whalers.

Hussey and Robinson, in "A Catalogue of Nantucket Whalers" (Nantucket, 1876), publish a list of the names of one hundred and forty-nine Nantucket captains who *commanded* British whaleships prior to 1812. When we consider that most of these captains had been captured with their ships and crews, and that the crews were also impressed, we get an amazing total of Americans in the British Fishery.

Scoresby ignores the above facts in his otherwise admirable work. When he refers to Nantucket, as he does on page 84, he states that several families "from Nantucket, near Halifax, in North America," emigrated to Dunkirk, France, by way of "enhancing the trade." He overlooks the fact that large numbers of the neutral Nantucketers were induced by the British to emigrate to and establish a whaling colony called Dartmouth, in Nova Scotia, and also to settle in Milford Haven on the Coast of England. (Macpherson, "Annals of Commerce," vol. 4, p. 347.) With the establishment of these colonies and the influx of new blood, England found her whale fishery for the first time well established as a paying venture. An article by Thomas Beale, printed in the *Quarterly Review* in 1839, states: "Our success now [1786] equalled that of the American Whalers."

England had paid in bounties to her whaleship owners between the years 1750 and 1769 the enormous sum of £613,261. (Scoresby, p. 79.) She was still paying bounties at the date of Scoresby's book (1820).

The Dutch Fishery was obviously on the wane by 1770, and by 1812 the fleet contained not much over sixty sail.

Scoresby explains the passing of the Dutch whalemen in the following paragraph (p. 113):

It has been seen that, in point of ability for conducting the whalefishery, the British, in their early attempts, excepting a few of the voyages of the Russia Company's ships, were universally eclipsed by the Dutch, and that, notwithstanding the English led the way to the haunts of the whale in the northern regions, and set the example of capturing this animal as an occupation; yet their labors were attended with such ill success, and their exertions were in consequence so much relaxed, that instead of becoming experienced in the trade, they soon lost the little advance they had made in the art, while they were under the direction of the Biscayans; and, in

THE GREENLAND FISHERY

consequence, were long under the necessity of hiring a great number of foreigners to assist them in the fishery. This obligation of the British to employ the Dutch fishing officers in their ships was probably the occasion of a popular mistake, that the Dutch were the first whale-fishers at Spitsbergen. But after the bounty system had been established a few years, the British became as expert in the fishery as the Dutch, and the two rival nations probably exercised an equal talent for many years afterwards. The talent for the whale fishery among the Dutch, however, was on the decline, and in consequence of the imitation of their manner by the British in the middle, and indeed so late as the ninth decade of the last century, the energies of the fishermen were never brought into action. The Dutch, from indulging a habit of coolness, became inactive, and the British too closely copied their example. *About the close, however, of the Century, two or three captains of the whale-fishing ships, men of abilities, commenced a system of activity and perseverance, which was followed by the most brilliant result. Instead of being contented with two or three large fish, and considering five or six a great cargo, they set the example of doubling or trebling the latter quantity and were only contented, so far as to relax their exertions, when their ships could contain no more.*

It is possible that in the final lines of this paragraph Scoresby pays his sole tribute to the genius of the Nantucketer.

"The British whale fishery reached its greatest prosperity in 1815, when there were 164 whalers on the ocean." (A. Howard Clark, "Fishery Industries of the United States," p. 197.) The English fleet at this time probably totaled not less than two hundred and twenty-five sail, a part of the fleet being in port.

The first ship of any nation to enter the Pacific Ocean and capture whales was the British ship *Amelia* in 1788. This venture was so successful that in the next few years the fleets of all the world followed her example.

After 1830 England's fleet rapidly diminished. (Jenkins, p. 262.) In 1852 the total English tonnage was 16,113, and four ships only were at sea. In 1868 there were only thirty vessels from six whaling centers, and in that year, Hull, the first of all English ports, sent out her last ship.

The Scotch continued to send out whalers and to-day are actively engaged in the steam whale fishery.

CHAPTER IV
CAPE COD, LONG ISLAND, AND NANTUCKET

THE American whale fishery was initiated with the salving of stranded whales. These were Right Whales, the same variety taken by the Basque Fishery at an earlier period. Their feeding ground was close in shore. The New-Englander was quick to appreciate the commercial value of the prizes left at his door, and for many years, until whales were driven from our coasts, drift and stranded whales were a constant source of profit to the southern Massachusetts towns.

When Captain John Smith sailed for America in 1614, he carried a crown permit to fish for whales. The Charter of Massachusetts Bay Colony granted the privilege of taking "all fishes-royal, fishes, whales, balan, sturgeons and other fish." In those days, whales were so plentiful that they were easily taken by small boats rowing out from the shore. Cape Cod, Nantucket, and Long Island all commenced their whaling operations in this manner. Try-works were set up at convenient places alongshore. Whales were cut-in on the beach, and the blubber boiled. As they became fewer and more timid, it became the custom to hunt them with small sloops towing or carrying one or two whale boats, which stayed at sea long enough to capture one or two whales and remove the blubber, which was stowed in casks and brought home to be tried-out.

As whales were driven farther off-shore by constant fishing, it became necessary to build try-works on the deck, and to give the oil a first hurried boiling at sea.

When brought ashore afterwards, this had to be refined. It is probable that wood was used under the try-pots in the early days; later, the blubber scrap furnished the fuel. The sloops were small and voyages were short, a few weeks or a month at the most. But with the discovery that the Sperm Whale was also man's legitimate prey, grounds farther away were fished—Hatteras and Bermudas—and voyages were still further lengthened. Nantucket soon outstripped all competitors, and became the great center of the world's whaling industry.

Early explorers and travelers speak of an abundance of whales along the New England coast. Richard Mather describes "mighty whales spewing up water in the air like the smoke of a chimney, and of such incredible bigness that I will never wonder that the body of Jonah could be in the body of a whale." (Sabine, "Fisheries of the American Seas," p. 42.)

It is said that the Indians first taught the colonists how to take whales. Waymouth's "Journal of a Voyage to America," in 1605, describes the Indian method of whaling in canoes; of first "getting fast" with a bone harping-iron having a bark rope attached, and finally killing the whale with arrows. Other contemporary writers explain that a "drug" was attached to the end of this rope.

These whales attacked by the Indians were probably blackfish, small cetacea of about twenty-five feet average adult length, but in exceptional specimens reaching thirty or even thirty-five feet. The blackfish schools have always frequented the waters about Cape Cod, and to this day the citizens of the Cape towns participate in a blackfish drive every once in so often, sometimes driving ashore and capturing several hundred in a single shoal.

The hardy and adaptable New-Englander, not content with the occasional drift whale that fell to his lot, quickly adapted the Indian method of fishing to suit his own requirements. He evolved a lance with which to kill the whale and early in the 1600's we find him busily engaged in the shore fishery.

The Cape-Codder was the first to attempt this fishery sporadically. The Long-Islander deserves the full credit for first making it an organized business. But to the Nantucketer will always be the glory of having founded a great national industry.

Little evidence is available concerning the early whale fishery of Plymouth. Almost from the inception of the colony there were frequent casual references to the whale and its products in the letters, papers, and books of the day, indicative that the pursuit at that time was of importance. But the documentary evidence was largely limited to court records of litigation over drift whales. The large and increasing number of these presupposes the agency of man in killing them.

The earliest records of organized whaling come from eastern Long Island. Here before the middle of the seventeenth century drift whales were apportioned by law, whaling companies were formed, neighborhood lookout stations were posted, and small craft put out from shore on cruises that lasted one or two weeks, making shore each night. (Starbuck, p. 10.)

To understand the high degree of specialization which Nantucket brought to bear on the whale fishery, far outstripping for a century or more all competitors, it is necessary to study the nature of her inhabitants and her physical geography. Her first settlers were religious refugees, forced out of Massachusetts by the intolerance of the Puritans. At Nantucket they found freedom, and practically no means of subsistence except from the sea. The island is of sand, with at best but an inch or two of top-soil. They had to fish or starve. Some of their fish was bartered on the mainland for such other produce as they required. For a while after the settlement they reaped a rich harvest of drift whales, a considerable proportion of which probably had been harpooned and "drugged" by the active Cape Cod whale fishermen. But about 1672, according to Macy, a Right Whale swam into the harbor and spouted in full view of the town for several days. The sporting instinct of the Nantucketer was aroused, and possibly his cupidity. A harpoon having been fashioned by a local smith, a boat put out, and the whale was taken. By this act was the destiny of the Island decided.

A committee of the townspeople, having carefully studied the economic situation of their island, concluded that the pursuance of the whale fishery was the best living open to the inhabitants. They decided to put to sea and hunt the whale in his native element. Overtures were made to James Loper, of Easthampton, Long Island, to come to Nantucket and teach the methods of capture then practiced. There is no evidence to show that Loper came, and it is believed the arrangement was not concluded. But the Nantucketers went ahead on their own initiative, and had fair success. In 1690, however, finding that the Cape-Codders were more proficient whalemen, they engaged Ichabod Paddock, and he crossed from the mainland with his family and became the real father of the Nantucket whale fishery.

While whales were plentiful, several lookout masts were maintained along the eastern and southern shores of Nantucket, and boats put off directly whales were sighted. Captured whales were towed ashore and at high water were beached and the blubber removed and carted to town, where several try-houses were maintained. The Right Whale frequented the New England coast from October until the beginning of June, and the humpback was present during the summer and fall months.

Eventually, from incessant hunting, the near-by herds were depleted. In 1760 the shore fishery was abandoned. (Macy, p. 31.) At its height, 1726, eighty-six whales were taken in one season, and the record day's catch was eleven. (Cheever.)

Some time about 1700 a drift Sperm Whale grounded on the Nantucket beach. The superior quality of its oil was noted, and when in 1712 Christopher Hussey was blown off-shore and encountered a whale of the same species, he promptly got fast, in spite of the storm which was raging, and having killed his whale he rode out the storm in the lee of its carcass and then towed it home. This was a milestone in Nantucket history. Immediately a number of sloops of about thirty tons each were fitted out, and the first Sperm Whale Fishery in the world's history was under way. These sloops outfitted for cruises of perhaps six weeks, and carried casks in which to stow their blubber. In 1715 there were six sloops Sperm whaling out of Nantucket; in 1765 there were one hundred and one sloops, brigs, and schooners; in 1775, over one hundred and fifty all told.

Until 1745 the market for Nantucket oil had been Boston. The Boston merchants forwarded the bulk of the commodity to London. The Nantucket merchants in that year decided to do their own merchandising, and to that end sent one small vessel with a cargo of oil to London. The venture was very successful, and thereafter they marketed their own produce in their own bottoms. On the return voyages the vessels brought back supplies required on the Island.

There is no finer example in history of communal enterprise than the Nantucket Whale Fishery. The inhabitants were uniquely situated for united effort. Geographically they were cut off from the rest of the world. The original settlers were Quakers, unusually considerate of the rights of others. Through intermarriage they were generally related to one another, and in fact were more like a large family than a civic community. Of more than average intelligence, hard-working and thrifty, the people were so law-abiding that there was little or no government in evidence on the Island. There were no paupers and no criminals, and, to indicate the simplicity of their dealings with one another, it is only necessary to say that there was not even a lawyer on the Island.

Most Nantucket boys started to learn either the cooper's, the boat-builder's, or the smith's trade at the age of twelve, and at the age of fourteen went to sea, according to St. John ("Letters of an American Farmer," London, 1782), to become petty officers in a few years. Generally they left the sea at the early age of forty or thereabouts, to attend to the shore end of the business, leaving the pursuit of the whale to the youth of the Island.

Every one on the Island was directly interested in the whale fishery. Each ship was owned in a large number of widely distributed shares, and the fruit of each man's labor was not limited to the wages he received. If he made harpoons, they must hold, for he owned a share in the ship which was to employ them. If he sold provisions, they must be good, for the content of the crew would contribute largely to the success of the voyage in which he was interested. If he made a whaleboat, it must be sound, for his own son might head the selfsame boat. The business transactions of the town were more like the affairs of a clearing-house than like the ordinary trade and barter of the average village. The amount of money in hand generally was small. The conduct of business was largely a matter of crediting one item against another. The result of an apparently losing voyage might actually mean income to the Island as a whole, since so many Islanders had profited in outfitting the ship.

It was this unity of interest and the keenness of spirit that the Islanders brought to the whale hunt that made Nantucket the greatest whaling port of her period.

Different and more remote whaling grounds were opened up successively. First the Bermudas, and the Grand Banks, then West Indies, Gulf of Mexico, the Cape Verde Islands; Davis Straits in 1732, Baffin's Bay in 1761, Gulf of St. Lawrence in 1763, Coast of Guinea in 1763, Western Islands in 1765, Coast of Brazil in 1774.

At first every encouragement had been given the infant Colonies by Great Britain to engage in this hazardous undertaking. But when some degree of success had been achieved, discriminative legislation soon put the colonial fisheries at a disadvantage. At first there were bounties in favor of British ships, and then for a while American ships also participated under certain restrictions. In 1761 duties were levied on oil and bone carried to England from the Colonies. By another oppressive Act of Parliament, Americans were not allowed to take their product to any other market, which practically compelled them to pay the English duties. In 1765, further legislation, limiting the waters in which they might whale and prohibiting their codfishing, resulted in partial failure for most of the fleet, for Nantucket whaling craft had been in the habit of combining the two ventures in their voyages.

Then came the Revolutionary War. Out of her fleet of a little over one hundred and fifty whalers, fifteen were lost at sea, and one hundred and thirty-four were captured and their officers and crews given the option of whaling under the English flag or of going to prison.

Most of the Nantucket whalemen were impressed into English service. Many of them continued to serve at the conclusion of the war. Many who had been imprisoned returned home broken in health, and many who had stayed at home found themselves, after nine years of inactivity, too old to take up whaling again. In the meantime, many families had been transplanted from Nantucket to start whaling centers in Halifax, Nova Scotia; in Dunkirk, France; and in Milford Haven, England.

In 1794, William Rotch, who had led the exodus from Nantucket to France during the war, returned to America and settled in New Bedford, having found himself unpopular with his townspeople, who considered him a deserter.

It will be seen that the blood of Nantucket was sadly depleted. When the War of 1812 broke out, Nantucket managed, with the assent of our Government, to make treaties of neutrality with Great Britain, and to keep some portion of her fleet at sea. For two years these ships confined their operations to humpbacking on the South Shoal. But there were so many strictures laid upon them that the Island came out of the War of 1812 slightly, if any, better off than she had been after the Revolution.

But by 1819 she had sixty-one whalers, and in 1821 she had eighty-seven at sea. As voyages became necessarily longer and as ships grew larger, the limitations of Nantucket's harbor became evident. The ever-shifting sands of the Sound caught many an unwary craft. No vessel of size dared to cross the harbor bar, except in daylight at flood tide. The Camels, a floating steam-propelled dock, was built in 1839 to overcome this natural handicap, and was used to float ships across the shoal harbor entrance. But the town saw the fishery being drawn to other centers because of more favorable economic conditions. In 1833 the Nantucket Fishery passed its peak, with close to one hundred ships. Although at an earlier period there were more keels, this was the year of her greatest tonnage. But the mainland town of New Bedford had already far outstripped her. After 1843, in which year eighty-eight vessels were registered, the number of Nantucket ships gradually diminished, until the last whaler, the bark *Oak*, crossed the bar in 1869.

CHAPTER V
NEW BEDFORD

STARBUCK states (p. 43): "In the vicinity of New Bedford, Whaling probably commenced but little prior to 1760." New Bedford and Fairhaven were at that time included in the township of Dartmouth. But in 1787, Fairhaven, Acushnet, and New Bedford were incorporated under the name Bedford. In 1812 the separate towns of Fairhaven and New Bedford were formed. Fairhaven at that time included Acushnet. In 1790 the village which afterwards became New Bedford had a population of about 700. By 1830 this had increased to 7592. In 1836 it was 11,113, and in 1855, the census nearest to the period of New Bedford's greatest maritime activity, the population was 20,389. The present population is about 120,000, of which more than fifty per cent is foreign-born.

In the "district of New Bedford" were a number of lesser ports, Westport, Dartmouth, Fairhaven, Wareham and Falmouth. Mattapoisett was the port of Rochester, until the two separated, about 1840. The old port of Sippican is now called Marion. Holmes's Hole has been renamed Vineyard Haven.

Daniel Ricketson ("History of New Bedford," p. 58) states that "It is well authenticated by the statements of several contemporaries, lately deceased, that Joseph Russell had pursued the business as early as the year 1755."

It is probable that whaling started from this vicinity somewhat before the date mentioned by Ricketson. Recently there has been found an original manuscript

THE WANDERER IN PORT, SEPTEMBER, 1920

SHIP NIGER, FULL AND BY
Painting owned by Mr. Lyman Delano

BARK STAFFORD OUTWARD BOUND
Owned by Mr. Milton H. Goshein

DRYING SAIL: BARK WANDERER, AUGUST, 1920
Painting owned by Mrs. F. Gilbert Hinsdale

log-book of a New Bedford whaling voyage in the year 1756. This is, so far as is known, the oldest existing log-book of an American whaleship. It is now in the possession of Mr. George H. Taber of Pittsburgh, a descendant of John Taber, *Master*.

Journal of a Whale Voyage
Sloop *Manufactor*, John Taber, *Master*

DARTMOUTH
[New Bedford at this time was a part of Dartmouth.]

6th day ye 9th of April 1756
weighed anchor in Acushnet River Bound to Nantucket and from thence on a whale Voyage. at 4 o'clock after noon anchored in Holms-Hole (Vineyard Haven) at 4 after midnight weighed anchor and at 12 arrivd at Nantucket being 7th day ye 10th and lifted our Rigging first day ye 11th instint went to meeting 2nd day ye 12th got up our Rigging and Shipd Zecheas Gardner on a 19th of the whales Clear of all Charges Except Small Stores 3rd day ye 13th Came out of Nantucket Harbor with Small Winds and variable at evening anchord under the point.

4th day ye 14th the wind being Southerly Stood for Holms-hole and anchord there at evening the tyde being spent

5th Day ye 15th the wind remaining Southerly in the morning Scraped and Slushed our mast and fild 3 casks of water and then beat up to Quickshole and anchord there

6th Day ye 16th (N.B. Spoke with Jonathan Coffen) the wind Shifting to East-ward after Some Debate put to Sea off of No-mans-land we lay by and Caught Codfish this day we had fresh winds and Rain at night got under Short Sail, myself very unwell

7th Day ye 17th Ran under a full Sail making our Course good yesterday and today SSW, wind at N.E Cloudy unpleasant weather with Same Rain at evening Brought to under Trysail

First day ye 18th being windy the wind at N.W. we lay under Trysail and Foresail. Still Cloudy.

2nd Day ye 19th 1756
Fresh N.W. winds and Some Clouds found a Dead Whale with one iron in her; But not worth Cutting in the afternoon lay our head to the West-ward under Try Sail and fore-Sail in   L 38 d — 13 m

3rd Day ye 20th was fresh N.N.W. winds we saw plenty of Sparmecitys and Struck four But Could not get one   L. 37 — 10.

4th Day the 21st was moderate S.W. winds Saw many whales and I struck one and got in a 2nd and a tow-iron after which Z. Gardner Lanc'd her and she Stove and over-set our Boat but did not hurt a man. Afterwards we got her But lost her head.

5th Day ye 22nd was Rough S.S.W winds So that we Could neither mend our Boat nor Cut our whale on Deack. Saw Whales.

6th Day ye 23rd was moderate wind at S.W. This Day we mended our Boat and Cutt our whale on Deak afterwards we found a whale which we cutt   L 37 — 14

7th Day ye 24th the fore part was moderate we Cutt our whale on Deack the latter part Rough we saw whales but could not try for them Spoke with
[bottom of page torn]

first Day ye 25th of April 1756 was fresh winds a Sharp Sea from the S.W. the wind Shifted from the S.W. to the N.N.W Spoke with a Snow from Glasco

2nd Day ye 26th we had fresh winds from the N.E. Cruisd in the forenoon in the afternoon lay-by Spoke with Thomas & Beriah Giles   Lat 37 d — 10 m N.

3rd Day the 27th was Rough whale weather we Killd a whale and took her on Board Saw one dead whale which we did not Cutt wind at S.S.W.

4th Day ye 28th we had Small winds but a very bad Sea from the S.W the wind N.E. we cutt our whale on Deack and now have our Hogsheads full and are very Desirous to fill our Small Casks. Spoke with Jonathan Macy and one Freeman   Lat 37 — 39 m N.

5th Day ye 29th was moderate weather the wind N we cruesd of and on But saw no whales   Lat 37 d — 30 m N.

6th Day ye 30th was Rough whale weather wind at S.W. we cruesed of and saw plenty of Whales But Could not Strike. Lat 37 — 13

7th Day ye first of May was rough weather wind at S.W . . . [several words missing] that we lay Bye . .

First Day ye 2nd of May 1756
was Rough weather and a very Large Sea from the S.W. last night was very Squally with Showers of Rain at 12 o'clock of noon we put up our helm for home to Gratifie the Crew wind S.W Course N. b E

2nd Day ye 3rd
was moderate weather a Small Breaz at S.W. We continued our Course N. b E. till noon were in Lat 40 — 2 then Souned had 29 fathom Course gravel which I Supose to be off the western shore then Bore away N.E.

NEW BEDFORD

The Sloop *Manufactor* to John Taber		will loe to 1 gallon of Roum	1/2/6
3/4 bushel peas	£ 1/7/0	to 2 lb tobacco	0/5/0
1/4 quintle codfish	1/9/0	Moses Quoneison	
to port charges	3/3/9	to 1/2 lb Chocolate	5
1 bushel Corn	1/0/0	to 1/2 tobacco	

JOHN TABER

This is a singularly complete document. There are less than a dozen missing words. The way in which voyages of the period combined fishing and whaling is clearly shown, and a vivid picture is given of the tremendous shoals of Sperm Whales which frequented near-by waters at the time. The cruising ground of the *Manufactor* was about midway between Nantucket and Bermuda. It may be taken that New Bedford was near the beginnings of her whalefishery, since the sloop went to Nantucket for its boat-header, but there is no indication that whaling was a novelty to John Taber and his associates. The very matter-of-fact beginning of the log is ample proof that this was by no means the first whaling voyage from the vicinity.

It is evident that the *Manufactor* had no chronometer, since longitude is not mentioned. Her easting and westing was figured by dead reckoning, that is, it was estimated from a record of the direction and distances sailed, and from the depth and nature of the bottom picked up by the lead.

The entry under May 19th, "lay our head to westward under Trysail and foresail," might be mistaken to indicate that the *Manufactor* was a schooner, but a sloop's staysail was always called a foresail at this period.

It would be interesting to know if the provisions listed at the end of the log constituted her entire food for the voyage. The item of chocolate is an especially interesting one. As the sloop appears to have carried but one boat, she probably had a crew of seven or eight men all told.

Joseph Russell, of New Bedford, had four sloops whaling in 1765 of about fifty tons each, which cruised as far as the Capes of Virginia after Sperm Whales. The Bahamas, Gulf of Mexico, Caribbean, Western Islands, Coast of Guinea, Cape Verde, and Coast of Africa Grounds were all discovered and fished before the Revolutionary War.

In 1770 there were many vessels of large tonnage from New Bedford, among them the brig *No Duty on Tea*.

When the Revolutionary War began, New Bedford's fleet numbered between forty and fifty whalers. (Ricketson, p. 71.) During the war they did not attempt to fish, but in spite of this precaution the New Bedford ships fared little better than those from Nantucket.

American privateersmen brought many prizes into the harbor during the earlier stages of the war. New Bedford was the only port north of the Chesapeake not in the hands of the British, and there was a rich accumulation of colonial stores of one sort and another. In September, 1778, Sir Henry Clinton, having been advised of this, sent an expedition against New Bedford by water. The ships anchored in Clark's Cove, landing about 4500 men by a boat bridge, who marched north into town. Houses were fired, and most of the warehouses and thirty-four out of the fleet of fifty whalers were burned.

The revival after the war was immediate. Ships laden with oil were at once dispatched to London. The *Bedford* of Nantucket was the first ship ever to fly the American flag in a British port. She arrived in London, February 3, 1783. According to the story, her cook, who was hunchbacked, went ashore, and was greeted by a British tar who slapped him on the back, saying, "Hello, Jack, what have you there?" to which he is said to have replied, "Bunker Hill, damn ye, will ye mount?"

The first whaleship ever to enter the Pacific was the British ship *Amelia*, manned by the Nantucket colonists of Milford Haven. This was in 1788. She returned in 1790, and word of the new grounds having been sent by the crew to their brethren in America, five ships were fitted immediately and sailed for the coast of Chili in 1791; four from Nantucket and one from New Bedford. The first American whaler to round the Horn was the *Beaver* of Nantucket in August, 1791, returning March 25, 1793. The first American whaler ever to fill ship in the Pacific was the *Rebecca* of New Bedford, Captain Joseph Kersey, Master, which sailed September, 1791, and arrived home February 23, 1793.

In 1800 the first lighthouse, a wooden structure, was erected at Clark's Point at the entrance to New Bedford Harbor. At the raising of this, which was attended by practically all the townspeople, a hundred-gallon try-potful of chowder was served to all hands and according to Ricketson's account, "No one became intoxicated on this occasion."

From an early date New Bedford built many of her own ships, and there were also shipyards in Mattapoisett, Fairhaven, and Dartmouth. The first vessel launched in New Bedford was the ship *Dartmouth*, built for Francis Rotch. She carried the first load of oil from New Bedford to London in 1767, and later was one of the ships concerned in the Boston Tea Party, on which occasion she was mobbed and her cargo thrown into the harbor. In those days people took their personal liberties so seriously that the manifest absurdity of fighting over such matters as drink and taxation never occurred to them.

A number of oil refineries were erected in New Bedford and also several candle factories. In the same way that Nantucket at an earlier date had been almost self-sustaining, New Bedford now manufactured most of the essentials on which her industry depended. Practically everything in the nature of stores and gear that went into a whaleship was made in the vicinity — ships, boats, sails, rigging, casks, whalecraft, biscuit, and salt beef. The origination of American "ready-to-wear" clothes is supposed to have resulted from the demands of the whaler's slop-chest, at the time when long voyages became the rule. At first only rough dungarees, but later street garments were fashioned in different sizes. This was done in the neighboring town of Rochester, by an enterprising tailor who made his patterns and distributed them among the women of his neighborhood, and later drove around to collect the completed garments.

At the conclusion of the War of 1812 many other ports of New England were encouraged to enter the whale fishery. The fleets of Nantucket and New Bedford were depleted, and the price of oil promised to remain high. A number of towns from New York State fitted ships, and Philadelphia, Pennsylvania, and Wilmington, Delaware, also tried the venture. The latter had five vessels and continued whaling for a matter of twelve years, but Philadelphia sent out only two voyages.

The ships of New London were very successful in their whaling ventures, and are responsible for the opening-up of a number of the most important whaling grounds. They also instituted the American Seal Fishery. This last was frequently carried on in combination with whaling, giving rise to the term "mixed voyage" which afterwards was applied to any voyage which took seals, sea elephants, polar bears, walrus, or any other animals in addition to whales.

Sag Harbor was settled in 1730 and engaged in shore whaling from the beginning of her settlement. Her first record of voyages is in 1760, when three sloops were fitted. Eventually she became the fourth whaling port of America, ranking next to New London. In 1846 she had sixty-three whalers. Provincetown with her large fleet of schooners and brigs ranked fifth. The prosperity of Stonington, Connecticut; Warren, Rhode Island; and Edgartown-on-the-Vineyard was firmly founded on sperm-oil. Sag Harbor had a near-by rival in Greenport, whose average fleet over a period of fourteen years, during the height of whaling, was ten sail.

It was not until the year 1818, according to Ricketson, "that the whalefishery in New Bedford was entered into with that spirit and perseverance which has brought it up to its present importance and elevated position in commercial pursuits." By 1823 New Bedford's fleet equaled Nantucket's, both in numbers and tonnage.

On page 40 is a table from Scammon, which shows the size and distribution of the American fleet in the year 1839, when more ports were engaged in whaling than at any other time. After this date the fishery began to center at New Bedford, and the smaller whaling ports one by one dropped from the list.

Little has been said of the part Fairhaven played in the whale fishery, but this is because her history is indissolubly bound with New Bedford's. They used the same

1839

Places where Owned	Ships and Barks	Brigs and Schooners	Tonnage
New Bedford	169	8	56,118
Fairhaven	43	1	13,274
Dartmouth	3		874
Westport	5	4	1,443
Wareham	2	2	904
Rochester	5	10	2,615
Nantucket	77	4	27,364
Edgartown	8		2,659
Holmes' Hole	3	1	1,180
Fall River	4	3	1,604
Somerset		1	137
Lynn	4		1,269
Newburyport	3		1,099
Plymouth	3		910
Salem	14		4,265
Boston		1	125
Dorchester	2		581
Falmouth	8		2,490
Provincetown		1	172
Portland	1		388
Wiscasset	1		380
Portsmouth	1		348
Newport	9	2	3,152
Bristol	5	1	1,752
Warren	18	3	6,075
Providence	3		1,086
New London	30	9	11,447
Stonington	7	5	2,912
Mystic	5	3	1,797
Sag Harbor	31		10,605
Greenport	4	1	1,414
New Suffolk	1		274
Jamesport	1		236
Bridgeport	3		913
New York	3		710
Hudson	8		2,902
Poughkeepsie	6		2,043
Coldspring Harbor	2		629
Wilmington	5		1,578
Newark	1		366

harbor, and the ships of the two towns docked indiscriminately on either side of the river wherever there happened to be a vacant berth. They hauled out more frequently in Fairhaven than in New Bedford. In 1852 fifty whalers were registered from Fairhaven, but this by no means indicates the extent of her interests; for her merchants owned shares in many New Bedford ships.

Although Boston was never one of the largest whaling ports, her total participation was considerable. There were only three towns in the history of the fishery that sent out ships over a greater period of years. Between 1784 and 1900 New Bedford had ships constantly at sea. Nantucket was represented eighty-five of these years, New London eighty, and Boston seventy-seven. At no time did Boston's fleet number over eleven sail, but her average was very constant and she was one of the last American ports ever to fit out a whaler.

In 1842 America had six hundred and fifty-two ships, and the fleets of all the rest of the world amounted to two hundred and thirty, a total of eight hundred and eighty-two.

The history of this period of the fishery is largely a record of the discovery of new whaling grounds, the descent of a large proportion of the entire fleet upon the newly discovered regions, and their consequent exhaustion.

Whales have their beaten tracks, and will always return to their accustomed haunts. Once a whaling ground was really "fished out," there was no chance of a sudden influx of whales from another locality. Grounds depended for repopulation on the propagation of the whales that had escaped extermination.

New whaling grounds were opened in the order of the table given below, and although many were fished for a considerable number of years, or even up to the end of the fishery, in general the discovery of a new ground drew away the bulk of the fleet from any previously popular cruising area:

Brazil Banks	1774	Coast of Japan	1820
Madagascar	1789	Zanzibar	1828
Coast of Chili	1790	Kodiak Grounds	1835
Off-Shore Grounds		Kamchatka (first bowhead ever	
Mid-Pacific	1818	taken)	1843
5° to 10° S.		Okhotsk Sea	1847
105° to 125° W.		Behring Sea and Arctic Ocean	1848

One of the central recruiting stations for the Pacific during the first part of the nineteenth century was Galapagos, where a whaleman's post-office was maintained. This consisted of a huge white-weathered turtle-shell placed on the top of a conspicuous rock. Each whaler deposited in the shell letters consigned to other ships, and overhauled the contents for letters addressed to her. The post-office still stands, but to-day it is a painted barrel on a post.

The Golden Age of whaling was between 1833 and 1863. During this period, due to the increasing number of ships and the growing scarcity of whales, the spirit of competition between ships was rife. There was no actual change in methods, but the tools and gear of whaling were very much improved and refined, and the toggle-iron was invented and generally adopted.

Although the American whale fishery passed its peak in 1846 with seven hundred and thirty-six vessels, New Bedford's fleet continued to wax greater until 1857, at which time she had three hundred and twenty-nine ships, barks, brigs, and schooners, a greater total than the combined fishing fleets of Boston and Gloucester to-day, and with three times the tonnage. The total American fleet of that year comprised five hundred and ninety-three vessels. The Arctic Bowhead Fishery commenced in 1848 and for fifty years was the most important branch of the industry.

The Gold Rush of 1849 seriously crippled the Pacific fleet. San Francisco Harbor had been an important recruiting station, but every vessel that called there in 1849 and 1850 was tied up, as the crews promptly deserted for the gold fields.

The Sandwich Islands were first visited by whalemen in 1819. Honolulu became the most important Pacific whaling port when the Gold Rush drove the ships from the Golden Gate. Whalemen, in fact, are held responsible for the very existence of this fine seaport town, since Honolulu was merely a native fishing village with a few grass huts before the whalers made it a port of call.

From 1818 the growth of the whale fishery was at a constant and steadily increasing rate: 16,000 tons in 1818, 55,000 in 1828, 125,000 in 1838. In 1846 it reached a tonnage of 233,000 for 736 vessels, the greatest in the history of the fishery. But, although this was the year of the greatest tonnage, it was not the year of greatest prosperity. The value of bone and oil increased for a number of years, and in 1857, when New Bedford alone owned 329 ships, her fleet brought home 48,108 barrels of sperm-oil, 127,362 barrels of whale-oil, and 1,359,850 pounds of whalebone (Ricketson). This was more than half the catch for the whole American fleet.

The introduction of gas as an illuminant in the early part of the nineteenth century caused much alarm among the whaling merchants. But it had little apparent effect on whaling. Possibly the price of Right Whale oil was depressed, but new uses and greater demand for bone, and the need for sperm-oil as a fine lubricant offset this.

The discovery of petroleum in 1859 marked the end of New England whaling, although it was some time before this was evident. New Bedford merchants at first laughed at the bare idea that such a sorry product as coal-oil could seriously threaten the supremacy of sperm.

The Civil War dealt the industry its first serious blow. Thirty-nine whalers called the Stone Fleet were loaded with New England granite and scuttled in an effort to bottle up Charleston Harbor. Seventy more were burned by Southern privateers. The "Whaleman's Shipping List and Merchants' Transcript" in 1862 published the following: "That Southern Pirate Semmes has already made frightful havoc with whaling vessels, and his piratical ship—the *Alabama*—threatens to become the scourge of the seas." The *Alabama* operated in the Atlantic Ocean, while the *Shenandoah* in the Pacific nearly wiped out the Arctic fleet.

At the conclusion of the Civil War the whale fishery was in no position to compete against petroleum. Undoubtedly if the merchants of New Bedford at this time had bent their united efforts to the support of the whale fishery, new markets might have been discovered, new methods, including the use of steam and explosives, would have been developed, and the history of the whale fishery would probably have been quite different in its final chapters. But the merchants of the town foresaw a losing fight, and began at once to divert their wealth and energies to building up another industry.

There was a short revival of whaling immediately after the Civil War, due to high prices, but in spite of the depreciation of currency at that time, the price of oil soon began to fall. Only the increasing demand for bone kept the fleet at sea.

In 1871, thirty-four ships were crushed in the ice at Point Belcher in the Arctic, and in 1876, twelve more were caught almost in the same spot.

In 1880, after several more disasters to the fleet, steam was first applied as an auxiliary to sails in the Arctic whalers. But ships continued to be nipped in the ice. The Atlantic Fishery had practically been abandoned, and New Bedford ships were outfitting on the west coast. At one time in the nineties, more vessels actually were sailing from San Francisco than from New Bedford, although the bulk of them were still Yankee-owned.

In 1924, for the first time in over two hundred years, after the wreck of the *Wanderer* and the return of the schooner *Margarett*, came a time when there was not a single Yankee whaler afloat on all the seven seas.

CHAPTER VI
THE WHALER

THE small vessels employed by the American whale fishermen, when they first began to follow whales off-shore, were sloops patterned on Dutch rather than English lines. The small harbors of our southern New England and Long Island towns were shallow and it was largely from necessity that light-draft Dutch rather than deep-draft English models were followed. The early settlers of this locality were familiar with the boats of both countries. They had as clear a heritage from the Dutch as from the English, since the Dutch had settled the south shore as far east as Block Island, and the Pilgrims had lived in Holland for a number of years before sailing for the New World. Reproductions of contemporary colonial prints in Clark's "History of Yachting" (New York, 1904) show Dutch vessels in American harbors. The *Sparrowhawk,* a Pilgrim boat of the first colonial decade, the hull of which is still preserved in Plymouth, has the characteristic draft and sheer of a Dutch sloop. Pinkies with high sterns which might suggest Dutch influence were built in New England as late as 1840. But a Mediterranean origin has been ascribed to these vessels. This is borne out by their after-hull construction which closely follows the French chebec. The French Canadian is probably responsible for their introduction. Many early New England vessels carried lateen sails, which were common at that time in the north as well as in the south of Europe.

The Dutch invented the gaffsail as early as 1500, according to the evidence of contemporary prints. The early Dutch gaff was short and straight and was hoisted by a single halyard. The sail was loosely laced to the mast. Some of the earliest Dutch prints appear to show mast-hoops, but these were probably loose rope-bands and not rigid wooden hoops. The use of a boom was common and the foot of a sail might be laced to the boom or merely clewed out to the end of it.

The invention of triangular headsails (staysails or jibs) generally is attributed to the Dutch. They are shown in signed etchings by Peter Bruegel, of Brussels, who died in 1568. Two of these are reproduced in "Les Etempes de Peter Bruegel l'Ancien," par René van Bastalaer (Brussels, 1908). Jibs, however, were at first used only on sloops.

The prototype of the schooner was a Dutch vessel similar to the sloop, which carried, *instead* of a jib, a small gaffsail set on a second and shorter mast chock up in the bows. This acted simply as a headsail. She carried no bowsprit, and the forward sail served exactly the same purpose as a jib, that is, it kept the head off. With the foremast in this position a jib would have been superfluous. The American "Block Island Boat," which has sometimes been termed a "two-masted cat-boat," is very similar to this early Dutch craft except that the foremast is longer in the American boat, her two masts being of about equal length.

The English gaffsail was in existence in the middle of the seventeenth century. It differed from the Dutch sail by being invariably loose-footed (it had no boom at the bottom), and, in furling, the sail was not lowered to deck, but was brailed or lashed to the mast, the gaff remaining aloft. The English gaff was much longer than the Dutch spar and was trimmed at the peak by two braces, called "vangs," which led to the quarter-rails.

Both England and America have claimed the invention of the schooner. The debated question has long been, Who first applied a bowsprit and triangular headsails to a two-masted fore-and-aft rigged vessel?

The name "schooner" is probably of Yankee coinage and our claim to the invention of the rig has been stoutly maintained largely on the strength of the name. A jib-headed two-master was built by Captain Andrew Robinson for use in the Grand Banks codfishery in 1713. The story has it that as she shot from the ways some one ejaculated, "See how she scoons!" It was shortly after this date that the name "scooner" began to appear in American records.

It would seem, however, that the Dutch are responsible for the development of the rig. The first volume of "The Mariner's Mirror" reproduces a painting attributed to the Dutch artist Van der Velde, which shows two schooners under British colors. Van der Velde died in 1707. Arnoldus Montanus, in his volume "America" (Amsterdam, 1671), shows a schooner in an illustration between pages 186 and 187. This is forty-two years earlier than the date claimed for the American invention.

What the Yankee did for the schooner rig was to adapt it to large seagoing craft. The demands of the West Indies trade required a vessel that could work rapidly to windward. Our prevailing Atlantic coastal winds are southwest, and a vessel southbound has to buck both a headwind and the Gulf Stream, or else hug a dangerous shore to avoid the latter. The schooners which were evolved frequently made the trip in half the time required of brigs of equal tonnage.

The gaff was lengthened and throat and peak halyards took the place of the single Dutch halyard; the foremast was stepped farther aft and its length increased until it was nearly or quite as long as the mainmast. At a later date fore-and-aft topsails were added. Square topsails were common well into the third quarter of the nineteenth century. The whaling schooner *Era* carried square topsails through the first decade of the twentieth century, being perhaps the last American merchant vessel to set them.

The English, until about 1750, called a small open schooner a "shallop" and a large decked one a "sloop." (See definition of Sloop in Blanckley's "A Naval Expositor" of that date.) The name "schooner" made its appearance in the first edition of Falconer's "Dictionary of the Marine" (London, 1769). Falconer's definition shows that the word and rig were well established at that time.

It is difficult to determine when schooners were introduced into the American whale fishery, since the name "sloop," even after 1713, was still commonly applied to all vessels with gaffsails, and the staysail of a sloop at that time was always called a foresail. Much of the confusion about the origin of the craft undoubtedly is due to this lack of distinguishing names. The story of the coinage of the name "schooner" can hardly be taken as evidence that the rig had not appeared in America previous to 1713.

The schooner has never been the general-utility rig of Great Britain, although many other European nations adopted it. On small vessels the British preferred the ketch rig; partly, no doubt, because it was British, but also because the ketch is snugger and has certain advantages of easier handling. If a larger vessel was required, in the rough waters of the British coast a brig was almost as handy as a schooner of equal size and was much abler.

A gaff rig will point into the wind closer than anything except a "Bermuda Boat" (an old rig which we have rechristened "Marconi Rig"). A schooner can outpoint a square-rigger or lateener and can be handled with a smaller crew; these are her two advantages. But she is a rough vessel in a seaway, for her center of wind pressure is too low when she is close-reefed. She is also an uncomfortable vessel in calm weather when a ground-swell is running, for there is nothing to prevent the sails from slatting across the deck. Moreover, there are difficulties attendant upon reefing the outboard part of a big mainsail that would surprise a shellback accustomed to squaresails.

At the present time the British ketch rig is rapidly being applied to our Massachusetts ground trawlers. It is called "the Western Halibut Rig." Since the introduction of power to the fishing fleet the large sail area of the schooner is proving a nuisance. The high wheelhouse, the straight stem, and the round towboat stern of the new fisherman leave something to be desired in the way of beauty, but she is a sturdy craft and bids fair to drive the Gloucester schooner from the seas.

Starbuck records the use of the schooner in the whale fishery as early as 1736, but it is probable that the rig was used soon after the first codfishing schooner was launched, since codfishing and whalefishing at that time were frequently carried on together in the same bottoms.

Clark, in his "History of Yachting," says that "before the Revolution, England allowed lumber to be imported from her American colonies in sloops only." This naturally led to the building of sloops of large tonnage. Vessels of this rig up to one hundred and fifty tons were not uncommon. Most of the whalers of the period 1750 to 1760 were large sloops. Macy (p. 52), speaking of Nantucket in 1756, says, "They began now to employ vessels of larger size, some of one hundred tons burden or more, and a few were square-rigged." Starbuck mentions several ship-rigged whalers previous to the Revolution, among them the *Africa* and *Bedford* of Nantucket. In 1782 St. John says (p. 162), "The vessels most proper for whalefishing are brigs of about one hundred and fifty tons burden." It is possible that some of the whalers listed as brigs were topsail schooners. In Starbuck's "History of the American Whale Fishery," the same vessel's name appears sometimes under both listings. Brigantines are occasionally mentioned. A brigantine of this period carried a fore-and-aft mainsail and square maintopsail and topgallant; a brig carried a square mainsail. The mainmast of the brigantine was generally loftier than the brig's. The records shortly after the Revolution show that the Nantucket and New Bedford fleets were almost entirely square-rigged, while Boston and Cape Cod had begun to specialize in schooners. The first bark to appear in the records is the *Hero* of New Bedford, in 1803. In 1819 another bark was added to the fleet, the *Gideon* of Nantucket. Nantucket in 1820 had seventy-two whalers, most of them square-rigged. (Starbuck, p. 95.) In 1830, four barks, fifty-nine ships, and three brigs sailed from New Bedford, while from Nantucket twenty-one ships sailed. But the bark rig soon proved its superiority in this branch of service. In 1858 more barks cleared from New Bedford than ships. Soon after the close of the Civil War in 1868, thirty-eight barks, four schooners, and four ships cleared from New Bedford. Provincetown that year sent out thirty-one schooners and two brigs; Nantucket, two barks only.

By this time the bark-rigged whaler had become so specialized to fit the requirements of the industry that she resembled a ship of another century more closely than she resembled contemporary merchantmen.

Whalers were fine ships, although fiction generally states the contrary. They were built to suffer the worst the elements could offer, including the crushing force of Arctic ice. The best material available went into their construction, and they were more strongly timbered than any other craft. I have never heard of a sagged or hogged whaleship.

Although they spent a greater proportion of their days upon the water than other vessels, the life of the average whaler was longer than that of ships of any other service. Due to the length of their protracted cruises, they were often sadly in need of paint, and canvas was patched just so long as it would hold together. The captain and officers were on "lays" and they did not propose to spend their own good money on sails while old ones would serve.

A number of whalers were built in New Bedford on clipper lines, and these had a fine turn for speed while their bottoms were clean, at the beginning of a voyage. It was the rule to carry studding-sails and also to set royals when "making a passage." But once a cruising ground was reached, royals were sent down to make way for the lookout. Quite as many whales are to be seen while cruising at four miles an hour as at ten, so a whaler was content to loaf along under easy sail except when "making a passage" between whaling grounds.

The big boats on the cranes, the heavy white-painted wooden davits, the square after-deck-house, the skids with upturned spare boats, the try-works forward, all conspired to give the whaler a tubby appearance which might altogether belie her lines.

It is probable that the fiction about the poor design of the whaler is partly the direct result of her longevity — she lasted so much longer than other vessels that, although she might have been architecturally abreast of her period when launched, when brought into comparison with the extreme clippers of the nineteenth century, she presented a tubby appearance.

The *Truelove* was launched in Philadelphia in 1764. She made seventy-two whaling voyages before 1868, under the British flag, having been captured by the British in the Revolutionary War. She was still at sea in 1873, when she revisited Philadelphia and was presented with a flag, as a testimonial to her one hundred and nine years of active service. Jenkins ("A History of the Whale Fisheries," London, 1921) gives a description of the *Truelove*. She was a ship of two hundred and ninety-six tons register, with a very pronounced "tumble home" (her topsides slanted inward). She had what he calls "pigsty bulwarks," that is to say, every alternate plank was out to allow water to run freely from the decks. It was a common American practice at the end of the eighteenth and the beginning of the nineteenth century to have no bulwarks between the quarter-deck and the forecastle. There were stanchions, topped by a rail, to keep the sailors from washing overboard, and this was called a "rough tree rail." Probably the *Truelove* was built this way, and afterwards was partially planked. A painting of the whaler *Ann Alexander*, made in 1807, shows the rough tree rail.

THE WHALER

There were other long-lived ships, among them the *Rousseau*, ninety-one years, and the *Maria*, ninety years. *The Charles W. Morgan* had eighty-five years of active service and was still sound when she was hauled ashore and turned into a museum. *The Triton* served seventy-nine years, the *Ocean* seventy-five years, the *James Arnold* seventy-four, the last thirty-two years under the Chilean flag.

It was not until the industry was on the wane that the important features of the different types of vessels were sifted out and combined, and the "typical whaler" evolved. For deep-sea work she had to be square-rigged. It was discovered that a bark was about as fast as a full-rigged ship and would lay-to much better: a very important feature, as a whaler had to lay-to whenever boats were lowered or hoisted. Moreover, she had to be handled at that time by not more than half a dozen ship-keepers, as the rest of the crew were engaged in catching whales. She must come-about repeatedly in order to follow the whaleboats in chase, and the bark rig was far more easily handled and was quicker in stays. One feature of the ship rig was retained on the mizzen; that was the crossjack yard (pronounced "crojek" by merchant sailors, "crotchet" by whalemen). To the ends of this dummy yard were led the mainbraces which trimmed the mainyard. This kept them free of the boats, and clear of the deckhouse. On her fore and main royal-masts, breast high above the crosstrees, were installed "hoops" for the lookout. In the Arctic a canvas shelter for the lookout, called a "crow's-nest," was erected, usually on the foretopmast crosstrees, but this was never used on a Sperm whaler. She retained the old spritsail yard and dolphin striker. The former had been generally discarded by the merchant service years before, and the latter largely went out of use with the advent of iron and steel ships. She retained hemp standing rigging, as it was more elastic than steel, a desirable quality in cutting-in. The huge cutting-blocks were kept suspended under the maintop, and the cutting-falls were rove when needed. "Preventer" pin rails were lashed in the main rigging several feet above the sheer poles, and running rigging was belayed there while cutting-in, to keep the decks clear.

Yards, lower masts, bowsprits, and the doublings of jibbooms and upper masts were generally painted white. I recall a few instances of black yards. There was much less bright work about a whaler than about a merchantman.

A whaler generally had no forecastle deck; there was a small forecastle head for the lookout at the knightheads. Anchors were not taken on deck, but were lashed to the bulwarks, one fluke inboard, secured to the lash rail. The arms overside were canted inward with the ring securely lashed to the cat-head.

Forward of the forecastle hatch was a huge windlass. In the early days when cutting-in and in getting up the anchor, even so late as the early fifties, this windlass had been turned over laboriously with handspikes by a crew of sixteen men. There was a double row of holes at each end of the long wooden cylinder, and eight men worked at each end, four forward and four aft. Turning the windlass in this manner was heartbreaking work. Sometime in the fifties a patent windlass with arms similar to pump brakes was introduced, and four or six men at either brake pumped continuously while a whale was cut-in or an anchor was catted. This was the method still employed when the last whaler made port. In recent years, a few large ships had auxiliary donkey engines to turn the windlass, but this was never the rule.

Many whalers were flush-decked, some had a quarter-deck, but it was generally a low one, often but one step above the main-deck. Immediately aft of the fore-hatch, so that the man tending fires stood upon the hatch-cover, was the try-works, two huge four-barrel iron try-pots (in the fifties there were sometimes three) set in a brick oven, with a water tank below the grates, to keep the fires from charring the deck. Hard against the after-side of the try-works was the work-bench. The space between the legs of the bench was floored and slatted, and used as a hencoop. Hogs sometimes roamed the deck, and if the captain wished fresh milk in his coffee, a goat also made the voyage.

The waist of a whaler was kept clear for cutting-in operations. Forward of the mizzen was a roofed structure called the forward house which carried the spare boats and frequently served as shelter for a vegetable bin. The roof was never characteristic of the Nantucket skids, so it is probable that this feature was not added until the sixties, although some presumably older photographs appear to show it. A square house filled the full width of the after-deck with a roofed-in alleyway cutting fore-and-aft through its center. There were no side alleys as in merchantmen, and this gave the whaler a high-pooped, galleon-like appearance which again suggested a Dutch prototype. The wheel was under this shelter and so was the after-end of the cabin skylight or companion. A whaler carried no binnacle, and the compass, a "double-card-steering" or "transparent" compass, was suspended on the inside of the after-skylight coaming. Directly below the compass was the captain's desk, where he could cast his eye upward and keep check on the course. The galley was generally in the forward end of the starboard side of the house. The after-end was a locker in which the ship's drudge, the second mate, kept his odds and ends. On the larboard side was the companionway and a small storeroom which served a variety of purposes. On deck forward of the house, to larboard, was lashed the blacksmith's chest of tools; to starboard was the cooper's. These chests had overlapping tarpaulin-covered lids, and rested on short skids.

If the captain's wife accompanied him on a voyage, a frequent occurrence, the ship was termed by the rest of the fleet a "Hen Frigate." The vegetable bin under the skids was usually turned over to her as a deck sitting-room, after the vegetables had been eaten.

Along the after-beam of the skids hung a row of mess-buckets, wooden-bailed;

along the forward beam hung a row of deck-buckets, rope-handled. Atop the skids, in addition to the spare boats, were extra poles for cutting-spades, harpoons, and lances, and also spare oars and spars for whaleboats and various boat-gear and craft. On the roof of the afterhouse were the line-tubs of the larboard and starboard boats. The waist and bow boats had little stagings suspended inboard, between the bearers and davits, to hold their tubs. A lighter structure similar to the forward house was frequently erected as a shelter for the try-works, and was called "the try-works' house." The roof of this was kept comparatively clear on account of the foresail which just cleared it. Lashed vertically to the mainmast were spare spars. Lashed to the foremast were spare poles. Generally, a spare topmast was secured to the larboard lashrail. The lash-rail was peculiar to whalers; often casks had to be secured on deck, so a stout rail was bolted fore and aft along the inside of the bulwarks.

In the starboard bulwarks amidships, just forward of the main rigging, was a wide gangway where a section of the bulwarks was removed for cutting-in. Above this, was carried, overside, a cutting-stage, composed of three planks and a handrail. This staging was lowered over the whale when in use, and lashed upright when idle. The whole of the midship section of the deck was sheathed with seven-eighths-inch pine. The cutting-spades wore this thin on a voyage, and in spite of frequent repairs it was always relaid for another voyage. Even within my memory, some of the whalers had painted gun-ports, a survival of the days when merchantmen hoped to scare away pirates with the subterfuge.

The after or captain's cabin was across the stern, between decks. The companion-way was aft on the larboard side behind the partition, and the captain's washroom was opposite. Forward of his washroom was his stateroom, and here in his stateroom was the one and only sybaritic touch aboard a whaler. His berth or bed was in gimbles, stone-weighted, so that as the ship heeled he remained on an even keel. So for a century the whaling-masters have rolled in a luxury that even the first-class cabin of a liner does not provide. There was always a sofa across the after-end of his cabin, where the skipper could snatch a cat-nap and still be within hearing of any disturbance on deck. Against the forward bulkhead of his cabin was his desk. The compass above it was equally visible from the main or forward cabin, since the bulkhead did not carry full height to the deck above.

In the main cabin was a table built around the mizzenmast housing. The table-top was provided with fiddles to keep dishes from going adrift in a sea-way. Either side of the table, fore and aft, was a long narrow bench with a hinged plank back. Between meals the back was pushed forward and the narrow passageway left clear. Down either side of the forward cabin were doors giving into the officers' staterooms. The pantry was generally reached through the forward bulkhead. The mate always had the after-larboard stateroom, the second and third officers generally occupied the forward one. There was a fourth officer to head the fourth boat. Sometimes he rated as boat-header, an experienced whaleman who had no ship's duties save lookout and cutting-stage and who outranked the mate; sometimes he rated as fourth mate, who stood little higher in the social scale than the boat-steerer of the mate's boat. His stateroom was generally forward of the captain's quarters. Occasionally there was another stateroom, and the cooper and steward and perhaps the cabin boy lived aft. Under a scuttle in the main cabin floor was the "run" or lazaretto, where delicacies for the cabin table, explosives, etc., were kept.

The term "steerage" did not have its usual significance on a whaler. It referred to the quarters of the boat-steerers, blacksmith, cooper, steward, carpenter, cook, and cabin-boy. The steerage had its own mess and a steerage boy to wait on table. It was entered through the booby-hatch, a name applied to any hatch-cover with a companion-slide on it. On a whaler this was the after-hatch, and the steerage was just forward of the cabin on the larboard side, but without any communication between the two.

The hold of a large whaleship was divided into an upper and lower deck, and was entirely given over to casks. When outward bound most of the casks in the upper deck were filled with provisions. The casks in the lower hold were filled with water; the ground-tier casks with salt water and the riders with fresh. There were several large metal tanks conveniently distributed in the upper hold, called "coolers." The freshly boiled oil was drawn into these and left until the temperature was low enough not to swell the wooden casks. Forward of the cabin on the starboard side, opposite the steerage, was the "sail-pen" where spare canvas and cordage were stored. A whaler always carried at least two complete suits of sails, sometimes three, and numerous extra storm sails besides. New ones were always headed up in casks.

The forecastle was a triangular-shaped room in the bows between decks. It was entered by a hatchway near the foremast. A double tier of berths went around all three sides, about eight berths to the side. There were no benches. When at sea the sailors' chests completely surrounded the room, two in front of each lower berth, and these served as seats.

Several three-masted schooners appeared in the whaling fleet in recent years. There were also a number of hermaphrodite brigs, and the barkentine rig was tried out on the steam whalers.

The English preceded us in the use of steam in the whale fishery. The first American steam whaler, the *Pioneer*, hailed from New London. She sailed to Davis Straits in 1866, and was crushed in the ice in 1867.

The first American steamer to go into the Arctic was the bark-rigged *Mary and Helen*, built by William Lewis, of New Bedford, in 1879. She sailed for the West Coast in 1880, and proved in every way a success. Her auxiliary engines could make about eight miles an hour, and in floe ice she could follow a lane to windward, and, if necessary, force her way through a closing lead. Her success led to a change in the character of the Arctic fleet. More steamers were built, most of them bark-rigged, averaging about four hundred tons each. At no time has their number exceeded a dozen.

There were several important developments of the whaler which were seized upon and applied to other ships. The first was the Spencer, a sail now obsolete. Spencers were gaffsails which took the places of the mizzen and main staysails. The Spencer made its appearance in the late eighteenth century, and was in general use until the flat cotton canvas of the clippers offset its advantages, and the old practice returned.

The second development, the boat-davit, is still in universal use. In the old days boats were hoisted with tackles leading from the yardarms or standing rigging, exactly as a Gloucesterman hoists her dories to-day. The whale fishery followed custom at first. But the exactions of the work made an easier arrangement necessary. J. A. van Oelen ["De seldsaame en noit gehoorde Walvischvangst," Leyden, 1684] gives early evidence of an effort in this direction. He shows in one of the plates an enormous beam just inside the taffrail which bridged the poopdeck and projected overside at each quarter a distance of four or five feet. For a century this beam was characteristic of English and Dutch whalers. The forward boat-tackle led to the mizzen rigging. When the boat was hoisted full height, a crane on the outside of the ship was swung out to support the bow, and the stern was lashed to the crossbeam. It became customary to suspend two boats in this manner at each quarter, one under the other. All other boats were stored on deck, or else towed astern. About 1800, it became the custom in the English Fishery to sling two additional boats, one at each waist. On low-sided vessels these were hung under beams similar to the one at the stern. They bridged the deck at a height sufficient to clear a man's head, and were supported on timberheads at either rail. On high-sided vessels a stationary crane or arm was used. A heavy upright was strapped to the outside of the bulwarks, and a horizontal arm was mortised to the top of the upright at the height of the mainrail. From the outer end of this the boat-tackle was suspended. A third piece was mortised to the other two, forming the hypothenuse of a right-angle triangle, of which the longer leg was the horizontal one.

Most English whalers were high-sided, English naval architecture still inclining toward "wooden walls." For this reason, when the boats were hoisted on these cranes with keels level with the deck, they were high enough for ordinary cruising purposes, although they were always taken on deck in the passages to and from the fishery.

But the Yankee, long before this, had invented the davit. The American whaler had been gradually evolved from a very small beginning. She had been first a sloop, then a brig or schooner, and finally a ship. As a result, she continued to be built close to the water. In order to give the boats sufficient clearance, the strong oak upright of the crane was carried high above the rail. The arm instead of being horizontal, was given a strong upward steeve; a short brace underneath gave it a triangular support and a davit was the result. This same davit without change has always been used in the fishery whenever a bent one failed, and had to be replaced at sea.

Steamed and bent wood davits appeared soon after 1800, as evidenced by the contemporary model of the *Deborah Gifford* in the New Bedford Museum, and by sketches in the log-books. Early scrimshaw teeth also show steamed davits, but scrimshaw was seldom dated. The bent davits were both stronger and springier. It was particularly for the latter quality that they were adopted, since it was essential that boats ride easily.

Iron davits have been tried in the whale fishery, but were found altogether too rigid for the purpose. White oak, eight inches square, proved to be the best material.

Whalers were generally designated among whalemen as "three," "four," or "five boat ships," according to the number of boats that were carried on the cranes ready to lower. A large ship carried a fifth boat on the starboard side forward of the gangway. Small barks like the *A. R. Tucker* and *Greyhound* could carry but two on the larboard side. Schooners and a few small square-riggers stowed their spare boats across the stern on "tail-feathers." But as a boat did not lower from this position, it did not count in the rating. The small early American sloops at first either towed their boats or carried them on deck. The American method of carrying boats on large ships in 1800 was the same as now, three on the larboard side and one to starboard.

About 1820, the English adopted the American method of slinging boats three in a row, doing away with the double-nested boats at the quarters. For a while they tried three boats on each side. Scoresby explains that "the harpooners having descended upon the whale, ... two boats, each of which is under the guidance of one or two boys, attend upon them, and serve to hold all their knives and other apparatus." This explains why the extra boats on the starboard side did not interfere with cutting-in.

Bunks or berths in the forecastle were introduced by whalemen. Sailors before the days of the northern whale fishery had always slept in hammocks. But in the Arctic winters a mattress was required to keep the sleeper warm, and no hammock constructed would hold a sailor and a mattress at the same time. A sailor's mattress — due to its straw or corn-husk base — has always been called "A Donkey's Breakfast."

The hull construction of whalers had to be the strongest. In the English Fishery, the ships were "fortified," that is, additional timbering was put in, and the bottoms reinforced. The Yankee ship was framed and built for her job. All her timbers were live-oak and her planking white oak. If she went into the Arctic, her bows were double-planked. Her bottom was always covered with seven-eighths-inch cedar sheathing, copper-covered for protection against worm and weed. Several whalers in 1830 were experimentally sheathed with leather.

In 1800, the value of a whaleship of approximately two hundred and fifty tons, outfitted and ready for sea, was between $10,000 and $15,000. In 1900, the value of the same ship would have been between $60,000 and $75,000.

Nantucket ships in 1820 averaged two hundred and eighty tons burden. While the Pacific was being opened and whales were plentiful, still bigger ships became the rule and were found economically suitable. The size of the vessels continued to increase until a considerable proportion of Nantucket and New Bedford ships in the early forties rated over four hundred tons. But under intensive fishing, the whales thinned out rapidly, or became more wary, and the length of voyages increased. Smaller vessels making shorter voyages could actually make as much money for the owners and on a much smaller investment. Moreover, there came a time when voyages got altogether too long; there was sickness and a breaking of morale, desertions followed, and a resulting moodiness among the officers not conducive to the success of voyages. The size of ships began to decrease. When New Bedford's fleet of square-riggers in the early 1900's had narrowed down to about a dozen, they averaged not more than two hundred and fifty tons each.

There have always been small vessels in the whaling fleet. The schooner *Olga* of fifty tons sailed regularly out of San Francisco in the early 1900's. The *A. R. Tucker*, of one hundred and thirty-eight tons, was broken up in the 1900's. In her day she was accounted the smallest bark-rigged vessel afloat.

The largest American whaler on record was the 807-ton ship *Sea* of Warren. She made one voyage with fair success, sailing in 1851 and returning in 1855. But there were very few whalers that exceeded four hundred and twenty-five tons. It was found that four active boats could take care of all the whales that were apt to be encountered on a voyage, and with that fact to work from, the practical size for a ship became a matter of simple arithmetic.

A great deal depended on luck, however, in a whaling voyage, and although the average length of voyages could be foretold with accuracy, individual voyages could be anything at all.

One of the most remarkable short voyages on record was that of the ninety-one-ton brig *Amaret* which in 1856 filled with whale-oil in twenty-one days. Her try-works were in continuous operation from the time they were lighted until the last barrel was bailed to the coolers. But this was after being frozen in the ice of Davis Straits for eight months. Captain William I. Shockley in 1901–02 filled the bark *Sunbeam* in thirteen months. His total catch was twenty-four hundred and forty barrels of sperm.

There is one widely current impression about a whaler that I feel it my duty to correct. A whaler does not always "stink to high heaven"—on occasion she is as sweet a ship as sails. To be sure, there is sometimes a slightly stale odor from the harness cask, when the beef is not exactly prime; and when boiling there is an unavoidable reek of burning scrap, since it would be impossible to carry enough fuel for trying-out purposes, even if it were not too expensive. This smell is unpleasant, and it is no wonder that a ship down wind is offended. But between boilings an American Whaler is clean as a whistle; her oil is all tightly coopered and there is nothing at all to take exception to.

The old British Greenland ships always brought home rotten blubber in casks, having no try-works aboard ship. Undoubtedly the British are responsible for the malodorous reputation of the Whaler. The smell of whale-oil is mildly objectionable. Spermaceti has a pleasant, even aromatic, fragrance. As for sperm-oil, I cannot smell it to-day without an attack of nostalgia; the faintest whiff and I see again the old New Bedford wharves, black with oil-soaked earth, rough-binned with seaweed-covered casks; and fringed with long rows of dismantled square-riggers, their jib-booms housed and yards cock-billed. Give me the smells of a spar-yard, a rigging-loft, or a New Bedford wharf and all the refinements of the perfumer's art hold not a single charm.

CHAPTER VII
THE WHALEBOAT

FROM the very beginning of the whale fishery, whaleboats of all nations have carried sails. The earliest prints of Spitsbergen show boats with both square and fore-and-aft sails. An etching by Van Luyken, "Whaling Scene near Nova Zembla, 1596," shows a boat with a gaffsail. A Dutch print of 1682 shows a jib and spritsail. Since the boat of the early days had a very shallow keel, the sail was for use only in a fair wind. Before "going on" a whale, the sail was lowered. In towing a dead whale to ship, when the breeze was fair, all the boats fell in a tandem with sails set; this is shown in a plate in Churchill's "Voyages" (1744, vol. I), and on the scrimshaw tooth reproduced herein.

American whalemen, in the late eighteenth or early nineteenth century, instituted the practice of "going on" Right and Bowhead whales with mainsail set. The earliest direct evidence I know of attacking a Sperm Whale under sail is a sketch of a whaling scene in the log-book of the ship *Washington* of Dartmouth, 1834–36 (in the collection of the New Bedford Whaling Museum). The American print, "The Chase," drawn by three New Bedford artists, Van Beest, Gifford, and Russell, in 1859, shows a boat "going on" a Sperm Whale with gaffsail set and oars manned. The original sketch for this picture, made on paper with exceedingly wide margins, was exhibited in a New Bedford shop-window in 1858 with a lettered invitation to all whaling *officers* to enter and write suggestions on the margin. The sketch was shortly withdrawn with the margins and back covered with comment. "The

Conflict," drawn by J. Cole, and published in New Bedford the same year, shows a "fast boat" with spritsail set. Captain William I. Shockley, of Dartmouth, informs me that his father, Humphrey A. Shockley, while captain of the ship *Two Brothers* of New Bedford, in 1840 went to the Crozettes and took twenty-two hundred barrels of whale-oil in eight months. All his whaling was done with paddles and oars. This was so unusual even at this early date, that it occasioned remark. It was doubtless due to the very windy nature of the grounds around these islands that sails were not hoisted. The several prints of the abandonment of whaleships in the Arctic in 1871 show boats with lugsails and spritsails, but no gaffsails. Captain Shockley tells me that the lugsail was quicker in stays than any other rig, and in the Arctic it was used almost exclusively during the seventies. The Sperm Fishery also employed this rig, but not to the same extent. An excellent model of a whaleboat, made by Captain Albert Robbins in the early seventies for the New Bedford Exhibit at the Paris Exposition, has a loose-footed gaffsail and a club-footed jib. The keel of the early boat for a short section amidships was rounded suddenly to a depth of about four inches. This served the double purpose of acting as a pivotal point when the long steering-oar was "leaned on," and of preventing leeway while under sail. The jib was always lowered before "going on" to give the boat-steerer room to dart his iron.

The bark *Morning Star*, Captain Henry D. Norton, carried a *centerboard* whaleboat on a voyage to the coast of Chile in 1857. This appears to be the first ever used in the fishery. The boat was equipped with a rudder. Neither Scammon (1871) nor Davis (1874) mentions the use of a centerboard, nor does Starbuck in 1878. J. Templeman Brown, in "Fishery Industries of the United States" (1887), describes a contemporary centerboard boat and gives the plans of one. Captain William I. Shockley used a centerboard boat in 1874, at which time he states "there were some in use." The centerboard appears to have been generally adopted by the New Bedford whalemen in the late seventies. With the advent of the centerboard, the sail area was increased, until in the late nineties the boom was fully twenty-five feet long, and the sail was peaked up until the gaff stood nearly parallel with the mast, and the peak far overtopped the masthead. With six men for live ballast it was easy to trim such a craft. A speed of over eight miles an hour has been claimed for the whaleboat under sail. The early mast was stepped through a three-inch hole in the second thwart, but the boat of the nineties had a hinged metal "partners" set in a carline which bridged the two forward thwarts; an inclined trough guided the foot of the mast into its step. After a whale was struck the rudder was unhung by pulling on an attached lanyard, which belayed to a cleat on the cuddy, and held the rudder securely slung outboard at the port quarter. There was no dead wood or skeg aft, as this would have slowed the boat in going about. In a light wind the men paddled under sail. Oars were not used, except in windward work when the noise would not carry to the whale.

THE WHALEBOAT

The whaleboat was without a rival as a surf boat. It might seem that because of the nature of its job a lifeboat would be better designed for this purpose. But this is not so. The lifeboat is a compromise boat, built to give the maximum capacity consistent with good surf qualities. The whaleboat, on the other hand, was the best sea-boat that man could evolve, with no limitations to size, weight, or model.

Every merchantman trading south of the Line carries a whaleboat for surf landing, since harbors are infrequent. The farther south one goes, the greater is the ground-swell; until, below both the Cape and the Horn, the ocean has a clean sweep around the circle of the globe, and the proportions of the swell at sea, and the surf on the shores of the South Sea Islands, are almost incredible to a Northerner.

I have sailed with a smart fair breeze up the gradual incline of a ground-swell until the whaleboat lost all headway and even gathered sternway. Then, when the crest had been topped, the boat slid down the long incline with all the exhilarating effect of a coasting adventure ashore. In such a sea it is good seamanship, even with a fair wind, to tack against the long swell so as not to lose steerageway. After the boats are once away, it is seldom that two catch sight of each other again. The sole indication of propinquity is gained from the masthead signals.

Even at the time of the early Spitsbergen Fishery, the typical whaleboat was a double-ender: that is to say, the stern was sharp and was framed in the same manner as the bow. But square-sterned boats also are frequently shown in prints of this early period, and these continue to appear in English prints even as late as 1781.

English and Dutch boats were much heavier than American boats, and their greatest beam was well forward. Six men has been the standard whaleboat crew in all nations, from the very inception of the industry. But it was not uncommon in the early Greenland Fishery to have five, seven, and even eight-man boats. All the early boats were clinker-built — that is to say, "lap-streak" or "strake," a construction which resembles clapboarding.

At the beginning of the nineteenth century, the typical British whaleboat of the Greenland Fishery was a carvel-built (flush-seamed) double-ender, with the bow straighter than the stern. [Scoresby, 1820.] The eight-man boat had disappeared entirely, the five-man boat was seldom seen, but the seven-man boat was still common, the extra man being called the "linesman." This man's duty is indicated by his name; he tended line. The line was not stowed in tubs, according to American practice, but in two square wells or compartments. The bollard (the English name for loggerhead) was in the bow of the boat.

The first American whaleboat was modelled after the Indian canoe. The early settlers had an opportunity to witness how handy a craft the canoe was, and they copied it closely, building themselves a clinker boat of thin cedar, about twenty feet long, with hollow entrance and run. It was flat-floored, and had its greatest beam amidships. This boat could easily be picked up and carried to the beach by two men.

The Honorable Paul Dudley, Chief Justice of Massachusetts, writes in 1725 ["An Essay upon the Natural History of Whales," Transactions, Philosophical Society of London, vol. III]:

> I would take notice of the Boats our Whale-men use in going from the Shoar after the Whale. They are made of Cedar Clapboards, and so very light that two Men can conveniently carry them, yet they are twenty Feet long, and carry Six Men, viz., the Harpooneer, in the forepart of the Boat, four Oar-men, and the Steersman. These boats run very swift, and by reason of their Lightness can be brought on and off.... Our People formerly used to kill the Whale near the Shore; but now they go off to sea in Sloops and Whale boats.

By 1782 the length of the whaleboat had increased to twenty-four feet. James Beetle, of New Bedford, built boats twenty-seven to twenty-eight feet long in 1827. In 1860, at the height of the Arctic Fishery, this length was increased to twenty-eight and twenty-nine feet, and in the nineties the thirty-foot whaleboat became the standard. Larger experimental boats have been tried out, but evidently without much success, since they were quickly given up. J. Templeman Brown ["United States Fisheries Report," 1887] mentions boats thirty-six and thirty-eight feet and six inches long, carrying seven and nine oars respectively. These were used for one season only, on the *Sally Ann* of New Bedford and the *Hannibal* of New London.

The American boat was introduced into the British Fishery at the beginning of the Revolutionary War, at which time most of the Nantucket fleet was captured and impressed into the British Fishery. Before this, the British prints show a rather blunt-nosed craft with the loggerhead in the bow. Boydell in 1789 published an aquatint by Dodd, entitled "The Greenland Fishery," which shows a typical American whaleboat with loggerhead in the stern. All the French prints show the American model, which is to be expected, since the Nantucketers established the French Fishery at Dunkirk in 1784, under Louis XVI.

The American boat has always carried six men. Macy says the *Beaver*, the first Nantucket ship in the Pacific in 1791, "carried seventeen men, manning three boats of five men each, which left two men, called ship-keepers, on board the ship when the boats were out in pursuit of whales." Macy is undoubtedly in error. The *Beaver* was a full-rigged ship of two hundred and forty tons, and two men would have been insufficient to handle her while her boats were away. Without question she lowered two six-man boats and her third boat was a spare. Early American whalers always carried a carpenter who was capable of making a new boat, if necessary, on the voyage, so one or two spare boats were sufficient at that time. In recent years, when good craftsmen were scarce, frequently as many as five spare boats were carried. Elisha Dexter, in his "Narrative of the Loss of the William and Joseph, Whaling Brig," Boston, 1842, writes: "Twenty-five years ago few vessels carried more than two boat crews ... the number of men to keep ship whilst the boats are out depends on the size of the ship. Thus a ship of three hundred tons (a three-boat ship) requires about five men to take care of her whilst her boats are in pursuit — in all twenty-three men and boys. A larger ship, a four-boat ship, requires thirty to thirty-two."

We have seen that the American whaleboat was clinker-built in 1725. The British boat was carvel-built in 1820, on account, Scoresby explains, of the greater strength required in rough work amid the ice-floes. The American boat was still clinker-built in 1834. [Hart, Miriam Coffin.] Beale describes an American clinker-boat in 1839. The Currier and Ives prints of 1852 also show clinker-boats. But the prints of 1859 and thereafter show only smooth-sided boats.

J. Templeman Brown, speaking of the "clinker"-boat, says that the name was formed in imitation of the sound made by the boat while going through the water. I have frequently noted this in a clinker-built tender. As the whale grew wary, the noise was found objectionable, and therefore a smooth-sided boat, to glide more silently upon the unsuspecting animal, was adopted.

The whaleboat, however, did not lose its old character. Its timbers were still of thin steamed and bent white oak; it was planked with white cedar, and its ceiling, platforms, and thwarts were of white pine. The planks were about one-half inch thick. Instead of being caulked like an ordinary carvel-built boat, the seams were reënforced on the inside by narrow battens a quarter-inch thick. The perfected boat may be called a composite clinker and carvel boat, since she retained three of her clinker-strakes. The second seam above her keel was still lapstreak, and so were her two top-side planks. The single lapstreak near her keel was retained in order to give the men a finger-hold in case of an upset. The beam of the boat was so considerable that they could not grasp the keel, and an easier reach was necessary. The upper of the two topside planks was not a true lapstreak; it resulted from a thicker plank which was employed to stiffen the gunwale.

A whaleboat had no deadwood aft, as this would interfere with quick turning. There was a very pronounced sheer and the "run" (after-body) was considerably finer than the "entrance" (forward end). An average boat was about twenty-six inches deep amidships; and rose to thirty-six to forty inches at the ends, the stem being slightly higher. The typical whaleboat was always "single-banked" — that is to say, there was but one oarsman to each thwart, and the men were "staggered," which means that the oarsman sat the full width of the boat away from his rowlock, instead of in the middle of the boat. This "staggering" of the men was necessary in order to balance the great length of the oars, which were the longest used in any service.

Thole-pins were true "thong-pins" in the British Fishery even so late as the time of Scoresby. A well-thrummed lanyard was attached to the end of a single fixed pin, and was hitched a short distance aft, to the gunwale. The thrust of the oar was against the pin, not against the "thole." The oars could not be boated readily with such a contrivance, so they were fitted with grommets at the handle, and when the boat was fast to a whale, the oars were thrust out to the grommets and towed alongside.

In American boats, double thole-pins were used until about 1850, when oar-locks were introduced. The Currier print, "Laying on" (1852), shows thole-pins, the Prang and Mayer print, "The Conflict" (1859), shows oarlocks.

The tub oar had a "crotch," a double-decked rowlock, about nine inches high, to lift it clear of the line-tub in a sea-way.

The oars in the Dutch and British boats were of uniform length, generally shorter than the American oar. The standard was from fourteen to sixteen feet in Scoresby's time. American oars have always been of graded lengths. This is explained by the difference of model in the two boats. The extreme width of the British boat was about five to five and a half feet. This width was carried nearly uniform to the bows and quarters, which were sharply rounded in. The boat sat much deeper in the water than the American boat, not being flat-floored. The American boat was much wider, measuring from five and a half to six and a half feet. The gunwales amidships immediately started to narrow toward the ends, after the manner of the Indian canoe. As a result, the beam of the boat varied greatly at the different thwarts.

The shortest modern American oar of which I can find record is fifteen feet, and the longest eighteen. The oars were always in three lengths, and the one long and the two short oars pulled on the starboard side against the two medium length oars on the port side. I will give here comparative tables of three boats. One is quoted from Scammon (1874); one is from a Provincetown schooner which I measured in 1904; and the third is the *Sunbeam's* larboard boat, in which I pulled tub oar in 1904:

	From Scammon	Provincetown Boat	Larboard Boat of *Sunbeam*
Length boat	28 to 30 feet	28 feet	30 feet
Beam	5½ to 6½	5½	6½
Harponeer and after oar	17	15	16
Tub and bow oar	17½	16	17
Midship oar	18	17	18

John R. Spears ["The Story of the New England Whalers," New York, 1910] states: "Long oars formerly used went out of fashion because it was found that they wore out the men in any but the shortest pulls." He further says that "the oar was formerly twelve to fourteen feet, but may now be no more than nine." Spears is in error in all his figures. The same mistakes have frequently appeared elsewhere. The length of oar has never been shortened in New England ships.

The standard steering oar was between twenty and twenty-three feet, and had a right-angle grip for the left hand pegged into the loom about one foot from the handle.

In the tropics, sticks called "spreaders" were crossed in the boats to keep the gunwales from warping or springing under the tightly lashed gripes.

Due to the extremely light construction of the American boat, it was not feasible to keep the heavy whaleline in her bottom while she was slung on the cranes, so the line-tub was invented, which was removed after the boat was hoisted. At first American boats carried a single large tub. But it was found to be more convenient to move two small tubs than one large one. One line-tub was on the larboard side between the after and tub thwarts, and the other on the starboard side between the tub and midship thwarts. Each held one hundred and fifty fathoms of line.

When the centerboard was introduced, the size of the after tub was increased to hold two hundred and twenty-five fathoms, and the smaller tub, which had to be squeezed into a space between the gunwale and the centerboard, held only seventy-five fathoms. As the whaleboat increased in length and beam, the waist tub grew, until it held about one hundred and twenty-five fathoms.

Previous to 1818, the bottoms of the boats were not painted; they were pitched with hot resin. [J. Templeman Brown, p. 240.] The gunwales are always shown painted, in the earliest prints, generally a dark color for the sake of trim. In recent years they have usually been painted black, occasionally green. I recall one ship with red "gunnelled" boats. White bottoms were always preferred for Bowhead Whaling, as this approximated the color of the ice, and white was also the most common color in the Sperm Fishery. But I have seen pale blue and pearl gray boats for this service. J. Templeman Brown says that the boats used in the Gulf Stream were sometimes painted a leather or salmon tint. Black boats have been used in the sea-elephant and seal fisheries and at times in whaling operations. A. H. Markham ["A Whaling Cruise to Baffin's Bay," London, 1875] states that it was the practice in the Greenland Fishery at that date for each ship to have her boats brilliantly painted in some distinctive manner. But, anything which would render the boat conspicuous, and so tend to gally the whale, was to be avoided. The use of the white boat was general during the last years of the fishery.

I have seen interiors painted lumber-wagon blue, light umber, salmon pink, and gunboat gray. These were the usual colors, and they were used either uniformly or in combination. The stern cuddy-board was generally painted the color of the outside of the boat. The "box" in the bow and the top of the thwarts were usually painted a different color, for trim. The ceiling was generally a light color, so that whalecraft would silhouette against it and be easily discernible. The short pine platforms at the bow and stern, where the harponeer and the boat-header stood, were *never* painted, as that would make slippery footing.

A whaleboat weighed from five hundred to six hundred pounds. About one thousand pounds of gear and craft went into her before the crew of six whalemen, weighing, perhaps, one thousand pounds more, were ready to take to the water.

Such was the Yankee whaleboat as it was finally evolved; the most perfect water craft that has ever floated.

THE CHASE OF THE BOWHEAD WHALE
Painting owned by Mr. George Peabody Gardner

SAILING DAY
Water-color sketch owned by Mr. Richard Sellers

THE FLURRY
Water-color sketch owned by Mr. Richard Sellers

LANCING A SPERM WHALE
In the collection of the New Bedford Public Library

SAMPLING OIL AND DRYING SAILS
Water-color sketch owned by Mr. Richard Sellers

COOPERING CASKS
Water-color sketch owned by Mr. Richard Sellers

CLOSE-HAULED

The bark Andrew Hicks, a five-boat ship of 303 tons built at Fairhaven in 1867. One of the last whalers to go out of commission
Painting owned by Mr. Edward T. Pierce, Jr.

JOURNEY'S END: THE WRECK OF THE WANDERER
Painting owned by Mrs. George Weymouth

CHAPTER VIII
THE WHALE

THE whale sought by the American colonists in their first ventures was the Right or Biscay Whale. Humpbacks were also to be found in great numbers on the near-by shoals, and although less valuable were undoubtedly hunted in the same period. The capture of the Right Whale was a comparatively simple matter, since the animal was sluggish and non-combative. The chief danger to be anticipated was a single sidewise swipe from his flukes when first struck. With reasonable care, safety to boats and crews was assured. In substantiation of this estimate of the relative tameness of the early shore fishery is the testimony of Macy, in his "History of Nantucket," that in the seventy years of shore whaling preceding the year 1760, when shore whaling ceased at Nantucket, not a single man lost his life in the fishery. This was in spite of the fact that there were several thousand men engaged in the pursuit. But this statement must also be taken as a tribute to the skill of the Nantucketer, for if a boat once got in the way of the flukes it spelled disaster.

The Humpback is a smaller whale than the Right, with pectoral fins nearly equal to one third his length. His capture was complicated by the fact that he often sank when killed; but since he was taken in shoal water it was not difficult to haul him near enough to the surface for towing, and sufficient gas generally formed in the carcass to float it in a very short while.

The taking of the Sperm Whale proved a different business. Not only was this whale well armed; but he was also naturally pugnacious. In the pursuit of his accustomed food his life was one of continual struggle, and he knew from experience exactly how to defend himself. Whale-fishermen of all nations had fought shy of him until the Nantucketers started in with their light boats, and at first using "drugs," familiarized themselves with his actions and devised the proper means of attack. If we read the comments in the right-hand columns of Starbuck's list of voyages, we find scarcely a page without record of violent death in the Sperm Whale hunt. The following items are typical: "Captain Brock and his boat-crew were lost while fast to a whale, Sept. 23." "Returned because Captain Maxfield's shoulder was broken by a whale." "First Mate taken down by a foul line." "Capt. Eddy died from injuries received from a whale." "Lost second and third mates and nine men." "Mate and boat-crew taken down by a whale." "Captain Harris and boat's crew lost, fast to a whale." "Lost, sunk by a whale." This last was the bark *Ann Alexander*, attacked by a fighting Sperm bull which had previously stove three boats.

There were four whales commonly taken by the Yankee whalemen: the Right, the Bowhead, the Humpback, and the Sperm. The Right Whale hunted at Nantucket and Long Island in the early days of shore whaling was the "Whalebone Whale" of temperate waters. The same whale is also found in the North Pacific Ocean and in the South Seas. Since the Right Whale does not cross the Equator, the whales of these three localities do not communicate with each other and they probably present slight differences, but these are not sufficient to constitute separate species. The name "Right Whale" is said to have originated in the early days from the whalemen's practice of singling out a particular whale, from among the more common finner whales which were difficult to capture, with the remark "There is the Right one."

The "Bowhead" is the great Arctic whale found only in the frigid waters of the North. The name "Bowhead" was applied by the Yankee whalemen when they first invaded the Pacific Arctic Ocean in 1848. In the early Spitsbergen Fishery this whale was called by the English the "Greenland Whale." The New England whalemen were introduced to him when the Nantucketers sent their first ships to Davis Straits in 1732.

These two whales, the Right and the Bowhead, are the only whales that produce whalebone of commercial length and quality, "size bone," as it was called, and for that reason they bear the inclusive name of "Whalebone Whales." They are distinguished from others by the total absence of dorsal fins or humps. Furthermore, they do not have the belly furrowing, called by the whalemen "Ginger rolls," which is characteristic of the other bone-bearing whales: the Finback, Sulphur-bottom, Humpback, and California Grey; the last named is also humpless.

The Right Whale has a comparatively straight snout with a "bonnet" at the nib-end. The bonnet is more prominent in the South Sea whales, but all Right Whales have it. This feature consists of a small pitted horny island about one or two feet across. It is generally infested with whale-lice, a variety of degenerate parasitic crustaceans.

The Right Whale is said to be black, but he probably is a dark gray with a tinge of actual color. All whales have a considerable play of color which changes with the light and dulls quickly after death. Scammon places the limit of the length of the Right Whale at seventy feet; the average length is under fifty feet. His head is about one fourth his total length.

The Bowhead of the North Polar seas, although perhaps shorter, is a much thicker whale. Scoresby gives his extreme length as sixty-five feet. Scoresby, who was a veteran whaleman, assisted in the capture of over three hundred of these whales, and his books are the most authoritative writings we have on the Greenland Fishery. The head of the Bowhead Whale is sometimes equal to two fifths of his length; generally it is about one third. His snout has a pronounced arch, which is responsible for the name "Bowhead," and there is no "bonnet" on its nib-end. His bone is longer and much finer in quality than that of the Right Whale, and due to the requirements of the colder waters which he inhabits, his blubber is thicker. As a result, he gives the greatest amount of oil produced by any whale, as much as two hundred and seventy-five barrels, or, it has been claimed, on occasion over three hundred barrels. His color is said to be a dark brown, but this will be found to be too dull an estimate except for a dead whale. Undoubtedly, whales of different localities vary somewhat in size, color, and proportions, due in part to local conditions of feed and water. For whales are not aimless wanderers; they have their accustomed haunts which they visit at definite seasons and they travel beaten tracks like other animals.

The Right and the Bowhead, the two Whalebone Whales, have many points of similarity. The Bowhead is more timid and the Right Whale more agile. Their habits of spouting and feeding are similar; once fastened to they will "sweep" their tails from "eye to eye" and will fluke and run in much the same manner. They both have a trick of "hollowing the back," which is most disconcerting to whalemen. But the Bowhead is much the harder whale to capture, for he is extremely difficult to approach and whalemen must work fast to guard against his escape under the ice.

Whales are mammals. They inhabit the water, but, unlike fish, they have no gills, and so have to come to the surface to breathe. They conceive and bring forth their young exactly as do other mammals, usually one at a time, and they also suckle them. The outward characteristics, distinguishing them from fish, are the absence of gills, fins, and scales. Instead of fins, they have muscular flippers, the flippers containing bones not dissimilar to the bones of the human hand. Their tails or flukes are horizontal instead of vertical.

The manner in which a whale breathes is perhaps the most interesting of his several specializations. His nostrils are at the back of his head, or, in the case of the Sperm Whale, at the top of the forehead. A whale stays under the water for a considerable period, and when he comes to the surface he projects his spout-hole and breathes. The breath is visible, and for that reason it has been termed a "spout," and for a long while was supposed to be water. All whales have double spout-holes except the Sperm Whale, whose spout-hole is single and shaped like an extended letter "S." It is situated at the top of his forehead, well to the left of the center. Although other whales have double spout-holes, they do not present double spouts, for the lines of the two spouts are parallel and the vapor combines as it leaves the head. It is said that either directly astern or directly ahead, the separation of the spouts of the Right and Bowhead Whales may be traced for a short distance. The spouts of all whales are vertical, except that of the Sperm, which inclines forward at an angle of about fifty degrees. The tallest spouts are those of the Sulphur-bottoms and Finbacks, which on occasion are said to be visible to a height of fifty feet.

Formerly the spout of a whale was supposed to be a jet of water. Even in modern times a few scientists have supported that opinion. Among whalemen, whose daily labors frequently took them into the midst of the spout, there does not appear to have been any diversity of opinion. From Scoresby on, they have accurately described its nature — a breath of damp air, visible because the moisture in it is condensed.

Nevertheless, there has always been a diversity of opinion as to just why the spout should be visible, that is, just why the moisture in it is condensed. We all know that the breath of a terrestrial animal is visible in cold weather when the moisture it contains meets the chilly air. But it is not visible in summer except when the animal sneezes. A whale's breath is visible the year around under all climatic conditions.

Roy Andrews ["Whale Hunting with Gun and Camera," New York, 1916, p. 42], speaking of Finback fishing in temperate waters, says, "When the animal comes to the surface, the breath that has been contained in the lungs under pressure is highly heated, and as it is forcibly expelled into the colder outer air, it condenses, forming a column of steam or vapor."

The descent into the compression of great depths would tend to send up a whale's temperature, but nature has provided all other animals with the means of keeping temperature constant, and there is no reason to believe that the whale is not similarly equipped, or that he could possibly continue to live if his temperature varied any considerable degree. It would appear that the temperature of whales is slightly, if any, higher than that of the human species. Guldberg (1900) and Racovitza (1904) say that the temperature of whales is inferior to mankind. Scoresby (1820) gives the temperature of a Bowhead Whale as 102°. The Sperm Whale's spout is visible under the tropic sun where the temperature of the outer air may equal or even exceed the temperature of the whale.

Racovitza, in "A Summary of General Observations on the Spouting and Movements of Whales" (Washington, 1904), states, "All gases under pressure which are suddenly liberated undergo an instantaneous reduction of temperature—in the tropics it is certain that the condensation of vapor is due solely to refrigeration caused by restraint."

The Sperm Whale feeds in a depth of not less than one half mile. All the known and charted sperm-whale feeding grounds are "off soundings." His food is the giant squid, which is a deep-water creature. A Sperm Whale in his flurry will sometimes vomit pieces of squid which are half the size of a whaleboat. That a whale goes down to such a depth is further substantiated by the testimony of many whalemen, who have had whales take lines out of their boats, sometimes to a length of almost a mile, in an approximately perpendicular direction. After remaining stationary for an hour or even longer, they have come up again within a few yards of the boat. This is a fact of common observation and not to be seriously questioned. It is evident that the initial spout of a whale upon arrival at the surface, after sounding to such a depth, will consist of air which has been subjected to terrific pressure, and this may be emitted with considerable force. Yet, although this first spout is larger and longer and more forceful than his second spout, it is proportionately no more visible, and it is not until the whale has been on the surface a considerable time that a gradual diminution of the visibility of the spout may be observed. This may be attributed to decreased volume and not to decrease in density. The whale inhales immediately after his first spout, showing that his lungs are relieved and indicating that no remarkable pressure is behind the second and subsequent spouts. A large Sperm Whale will spout fifty to sixty times before sounding. The final spout is clearly visible. The exponents of the pressure theory would have us believe that there is still sufficient pressure behind a spout to account for its visibility after a whale has been at the surface for fifty minutes and has breathed fifty times.

Racovitza further states that "The proof that the air is expelled under strong pressure is that the spout rises to a very great height, and especially that its emission produces a harsh sound, so characteristic that all authors have compared it to the escape of steam under pressure."

The breath of a horse that has been driven rapidly in cold weather is visible for a distance of three feet. His breathing is very audible and expiration lasts about half a second. The height of a Sperm Whale's spout is not more than ten or twelve feet, and its duration averages three seconds. The spout-hole is from six to ten inches long and opens to a width of several inches. Photographs of other varieties of whales, showing wide-open spout-holes at the moment of breathing, are given by Dr. Frederick True ["On Some Photographs of Living Finback Whales from Newfoundland," Smithsonian, 1903] and by Roy C. Andrews ["Whale Hunting with Gun and

Camera," New York, 1916]. If there were forceful emission through spout-holes of the size shown, the duration of the spout would be very much shorter than has been recorded; the lungs would be emptied in a single blast.

It is known that the inspiration of all whales takes very much less time than the expiration. Beale and Racovitza are agreed that the spout of a Sperm Whale takes three seconds and the inspiration one second. It is inconceivable, if any considerable pressure is required to empty the lungs in three seconds, that the whale could fill his lungs three times as rapidly, even if the spout-hole is capable of greater distension than is employed in the act of spouting.

There is plenty of testimony to show that the ordinary spout of a whale is saturated with moisture. I have proved this in the whaleboat, having been well dampened in its spray. Dr. F. A. Lucas states that he once stood on a ship's "topgallant forecastle" and there was moisture enough in a Finback's spout — to windward of him — "to drift aboard and feel like a whiff of fog."

W. M. Davis ["Nimrod of the Sea," p. 185] makes the following statement: "But no whaleman has witnessed a jet of water coming from the spout-hole of a whale — *the very blood which clogs the lungs* after the death thrust is blown into the air as *a fiery spray or mist.*" I have seen this thing myself; liquid blood instantly converted into vapor. The action of the spout under those circumstances is exactly parallel to the action of an atomizer.

The breath of a whale undoubtedly contains a much greater proportion of moisture than the breath of any other mammal. This moisture has been forced into his lungs from his tissues while at a depth, and his visible spout is due to this excessive moisture and not to high pressure or high temperature, since neither of these is present while the whale is at the surface. The amount of moisture in a whale's spout is sufficient to cause its visibility. When blown into the outer air the moisture-saturated breath atomizes instantly.

Whale blubber is a substance which has no counterpart in the anatomy of terrestrial animals. It is not fat as we know it — that is, it is not reserve matter — but an actual organ of insulation against both cold and moisture. It is by no means soft, as is generally supposed. In fact, it forms an almost impenetrable cylindrical casing for the whale, offering great resistance to pressure and also to moisture, since blubber is about seventy-five per cent oil; and oil and water do not mix.

There is nothing else in nature to parallel the adaptability of the whale to varying pressure and temperature. The highest flying bird is subjected to only a fraction of the degree of change experienced by a whale in sounding. The feeding ground of the Sperm Whale is frequently in waters the known depth of which is close to a mile.

When a whale sounds, he does not adjust himself to pressure in the way a human "sand hog" does in an air lock. In the latter case, the man breathes through both skin and nostrils and the air under increasing pressure gradually penetrates into his tissues. If pressure were suddenly removed, the man would swell with the expanding air within and would suffer an attack of "bends."

Fishes conform to the pressure of the depth in which they live. They breathe with the aid of gills which separate the air from the water. The flimsiest fish may live in comfort at great depths, but if suddenly brought to the surface will burst. This is the amazing thing about a whale; he not only can but he does without inconvenience constantly undergo rapid changes in pressure that would kill any other known living thing.

There is no connection between a whale's breathing passage and his throat, nor is there any evidence that he swallows his food at the same depth in which he finds it. Since his blubber is impenetrable and he does not breathe while down, as fishes do, there is no reason for increase of internal pressure through the admission into his anatomy of outside fluid or gas, as is the case with other creatures when they are subjected to added outside pressure. So whatever resistance the whale summons against depth pressure comes from within — either by the formation of internal gases, by the contraction of his bulk, or by structural resistance.

The generally accepted explanation of the whale's ability to withstand outside pressure is that he "jacks up" his internal pressure; his heart speeds up, his temperature rises and as a result gases are formed in his blood and tissues sufficient to resist the weight of the water above him, and that these gases escape in the spout when he returns to the surface. Whales have sounded to a depth of six hundred fathoms or more in about four minutes and have come up again to the surface just as rapidly. We are asked to believe that gases can form and pass through muscular tissue at the rate which would be necessary for such an adjustment at each sounding. But all whales, including sperm, are liable to sink the instant they are killed even if they have just emerged after a long submersion, which proves they are not inflated.

Scoresby [p. 250, vol 2, "Arctic Regions and Northern Whale Fishery," Edinburgh, 1820] has computed that a large whale at a depth of eight hundred fathoms is subjected to a pressure of 211,200 tons, an amount about equal to the combined weight of the four largest passenger steamships now afloat (April, 1926 — *Leviathan* 59,957 tons, *Majestic* 56,551 tons, *Berengaria* 53,336 tons, *Olympic* 46,439 tons). If a whale should approach the surface with such an amount of internal pressure, he would either burst, or he would have to swell to allow the gases to expand sufficiently to equalize inside and outside pressure; that is, to a size equal to the displacement of these four steamships! But he does not approximate this. He comes to the surface and is barely awash, not appreciably higher than before he turned flukes and went down. He does not breathe before reaching the surface and the manner of his first spouts suggests nothing more than the natural puffing of an animal that has held his breath for a while and whose lungs require oxygen.

Right and Bowhead Whales feed at the surface. There is no good biological reason for them to sound deeply except to escape from man or to pass under ice. The Sperm Whale alone must get his food at great depths. This is borne out by the fact that unless attacked the Baleen Whales submerge for not more than ten or twelve minutes, but a large Sperm Whale usually stays down for about an hour.

When Right and Bowhead Whales, under attack, have sounded deeply and have stayed down longer than is their habit, they return to the surface obviously exhausted; so much so that sometimes they lie dazed upon the surface and are easily killed. But the Sperm Whale, who is accustomed to these depths, gives no indication of inconvenience if the length of his stay is prolonged.

The Sperm Whale's belly and flanks exhibit a curious longitudinally wrinkled surface. This is also observable on Right and Bowhead Whales, but to a much less degree. It is quite different from the accordion pleat-like furrowing on the bellies of Finbacks, Sulphur-Bottoms, and Humpbacks, which allows these animals to expand to a marked degree. When killed the latter will swell up like balloons as soon as decomposition sets in, but Sperm, Right, and Bowhead Whales, the "smooth-bellied whales," will retain more nearly their normal bulk until they burst with the force of gases inside. These wrinkles on the blubber of the smooth-bellied whales may be taken as evidence of the whale's ability to contract into much smaller compass when under pressure.

The Sperm Whale is much more sturdily built than other whales, and after a considerable contraction will offer great structural resistance. Right and Bowhead Whales do not sound so deeply. They are less compactly built than the Sperm Whale and the blubber wrinkles are much less pronounced.

The nature of the whale's spout, his uniform bulk upon sounding and upon returning to the surface, and the rapidity of his descent and ascent suggest that Sperm, Right, and Bowhead Whales have some agent of resistance other than gas. I believe that the smooth-bellied whale's ability to contract under pressure is the most important single factor in his adaptability to various depths. Such contraction, by increasing his specific gravity, would prove of great assistance in swimming at low levels. Voluntary contraction would also explain the Right Whale's habit of "settling." Without apparent movement in all his vast bulk and without leaving any trace of air bubbles on the surface, the Right Whale will when disturbed drop like a plummet far out of the reach of the whalemen who are attacking him. This habit is allowed for in attacking the creature. The boat rows or sails straight at his flank and, as he settles when struck, the boat does not slacken pace but passes completely over him.

I would not preclude the existence of other agencies of resistance, but I am convinced that contraction is the most important factor in the resistance of these three whales, and that structural resistance and the formation of gas may be considered auxiliary.

A whale's circulation is supposed to be proportionately very much stronger than a land animal's. His heart will speed up under pressure and the lack of fresh air. A certain amount of gas will form and will contribute to his resistance. This will eventually escape through the spout. On the other hand, considerable moisture probably is condensed by pressure while at a low depth and evaporates again as the whale comes to the surface.

The whale lives to be a very old animal and apparently reaches full size at a comparatively early age. A hundred-barrel sperm bull sixty-five feet long is a very large whale and an exceptional animal. Nowadays a whale this size will have teeth weighing at the extreme less than two pounds apiece, generally very much less. But in old examples of scrimshaw it is by no means uncommon to find teeth very much heavier. These large teeth can only be explained in two ways. In the days before the Sperm Whale herds were depleted, there must have been exceptional whales, either larger or older than are found to-day. The intensive pursuit of the Sperm Whale began about one hundred years ago, and for fifty years big whales were singled out for capture whenever a pod was sighted. In that fifty years, judging by the increase in time required to fill a ship, probably ninety per cent of all Sperm Whales were killed off, and since the big ones were especially sought and presented the bigger targets, it is to be presumed that very few whales alive at the beginning of that period lived to see the end of it. Fifty years from now whales with large teeth probably will again make their appearance. Whether they will be giants or merely very old whales is a matter for conjecture. Captain George Winslow, in the bark *Desdemona*, in the late seventies took a Sperm Whale off the River Plate, in which the two largest teeth were eleven inches long, the pair weighing eight pounds seven ounces. They are in the collection of the New Bedford Whaling Museum. This whale is reported to have been over ninety feet long. So far as is known, his teeth are the largest that ever were taken. Other teeth from the same whale were smaller, as the size of Sperm Whale teeth gradually diminishes toward the back and tip of the jaw, the largest pair being about one third back from the end. Badly worn teeth are frequently found and may also be taken as an indication of age, although the tooth of the Sperm Whale is comparatively soft. It has no enamel and when green is not difficult to cut with a knife. The food of the whale is soft, but some whales have the habit of snapping their jaws, "gnashing their teeth," and this habit, possibly due to an aching tooth, may account for excessive wear. Diseased teeth have occasionally been found. A large whale has as many as fifty-two teeth; a small whale has fewer. As a whale grows, additional teeth are formed at the back of the jaw. These are conical in form, curved slightly backward, and are six to eight inches apart from centers. About one third of the length is visible, the rest being buried in the gum. The teeth are not for mastication, but serve to chop squid into large chunks, which are swallowed whole. They are also the whale's chief weapon of offense. The lower jaw when closed fits into a long narrow depression in the snout, burying about one fourth of its bulk. There are no visible teeth on the upper jaw, although there are abortive ones. The lower teeth grind against a double row of hard disc-like depressions in the roof of the mouth.

The bull Sperm Whale is very much larger than his mate, in bulk probably three or four times greater, and his head is much longer in proportion, sometimes amounting to almost one third of his length. The cow whale's head is about one fourth of her length. The female Whalebone Whale is said to equal in size or even exceed the male. The disparity in size between the sexes of the two varieties is explained by the fact that the Whalebone Whale is monogamous and the Sperm Whale polygamous.

The fact that Whalebone Whales can swallow nothing larger than a herring has furnished occasion for many a skeptic to deny the whole of the Old Testament, and the fact that a Sperm Whale could have swallowed Jonah and his horse as well (if Jonah had a horse) has been quite generally overlooked. This is probably because Whalebone Whales, for the first one hundred and fifty years of the fishery, were the only whales captured. Although the Sperm Whale has so accommodating a throat, his mouth is by no means so large as the mouth of the Whalebone Whale. Scoresby has pointed out that on occasion the mouth of the Bowhead may be one third of his total length. Since the total length of the Bowhead is frequently as much as sixty feet, this gives a mouth twenty feet long. Baleen from the Bowhead Whale on two occasions at least has been taken over seventeen feet long. This would indicate a whale seventy feet long with a mouth large enough to hold a four-horse team.

The Whalebone Whale's method of feeding is to swim rapidly with his mouth wide open through a field of "brit," small red, shrimplike crustaceans that live near the surface. When his mouth is full, he closes it. This manner of feeding is termed by the whalemen "scooping." Instead of teeth, the upper jaw of the Whalebone Whale is fitted with some five to seven hundred thin tapering slabs of baleen, commercially called "whalebone," weighing sometimes as much as thirty-five hundred pounds in an exceptional Bowhead. These fit side by side flat against each other with a hair-like fringe on the inside edges. The baleen is really a continuation and a modification of the natural corrugation in the roof of the mouth. It is a growth similar to nails, hoofs, and horns, and is chemically about the same thing. When the whale's mouth has closed, he pushes his huge tongue forward, forcing out the water. The baleen acts as a sieve, and holds the brit ready to be swallowed. The mouth opening of a good-sized Bowhead, as he rushes through the water feeding, is not less than thirty feet. The huge tongue may try-out as much as twenty-five barrels of oil. The lips will yield sixty barrels more (in a two-hundred-and-fifty barrel whale). The tail of such a whale would be about twenty-four to twenty-six feet broad and five or six feet deep, according to Scoresby, and is considerably more crotched than the Sperm's. It is also a much more curved and shapely tail, the Sperm Whale's tail consisting of two almost triangular flukes.

The Bowhead Whale, except when he is feeding, seldom stays at the surface more than two or three minutes at a time, in which short period he spouts seven to ten times. He then rounds out (hunches his back out of the water) and turns flukes (lifts his tail in the air) preliminary to sounding, in the same way as does the Sperm Whale. He stays down, however, only five to fifteen minutes. In recent years he was hunted exclusively for bone. The oil, called "train oil" in the British fishery, was merely a by-product, being inferior in quality to sperm-oil. Frequently it was not saved, a very wasteful proceeding.

After being taken on deck the bone was separated, and when all flesh had been scraped off it was bound in bundles of about sixty slabs. Frequently it was lashed to the standing rigging with the butts resting on deck. When stowed below it had to be guarded against rats, dampness, and cockroaches.

Sperm Whales are found in pods or schools, except for an occasional "lone bull" whose life history is analogous to a rogue bull elephant's. But Bowhead Whales are not gregarious. Generally they swim singly or in twos and threes. Sometimes on the feeding grounds they are found "herded," but this is a utilitarian, not a social, arrangement.

When attacked, a Bowhead is apt to sound vertically. He goes to a considerable depth, sometimes as much as forty-eight hundred feet, where he remains "sulking" for a considerable time. When he returns to the surface again, he is usually very much exhausted. Scammon records an occasion when one stayed down an hour and twenty minutes. Often he buries himself in muddy bottom. The Bowhead is not so accustomed to these depths as the Sperm Whale, and it is probably for this reason that he imagines safety lies in this direction. If in the neighborhood of ice the Bowhead is very apt to make for it, dashing through the water with flukes lashing from side to side, and "bellowing frightfully" according to Scammon. When he dives under the ice, he is apt to escape, as the boat, unable to follow, must either cut line or permit it to run out and be lost.

Whales have no vocal cords but when in violent action, and sometimes in ordinary breathing, they make audible sounds. Captain Shockley says that only the Right Whale "bellows." The Bowhead and Humpback make a "singing" sound "in the lungs" which can sometimes be heard for a distance of over a mile. It is always preceded by violent action.

When gallied, both Right and Bowhead Whales have a trick of sagging or "hollowing" the back. This causes the blubber to become limp, and if an iron is darted at this "slack blubber" it will not penetrate. The iron will generally fall back into the water with the shank bent double.

Besides mankind the Whalebone Whales have one other formidable enemy called by whalemen the "Killer Whale"; a large variety of mammal of the dolphin family. Their generic name is Orca. The largest are about twenty-five feet long and have very powerful teeth on both upper and lower jaws, and very long and erect dorsal fins. They hunt in packs, and when attacking a whale they seize upon the under lips, dragging the head down and forcing open the mouth. Generally, they succeed in drowning the whale, after which they proceed to eat out the tongue, which appears to be the desired morsel. These are the only "sword-fish" that attack whales.

The blubber of the Bowhead Whale is very much thicker than the covering of the Sperm Whale. Probably its extreme thickness in large and fat specimens is in the neighborhood of two feet. The Sperm Whale's blubber very seldom reaches twelve inches at the thickest parts; the average is not more than six inches.

Whalemen believe that whales sleep on the surface of the water. In support of this they instance that whales, lying perfectly motionless on the surface at night, are often closely approached by ships. When finally disturbed the commotion they make is so panicky that it indicates they have been startled from a sound slumber. In daytime whales may run, but they do not appear frightened.

The Sperm Whale is a more methodical creature than other varieties, and whalemen have a rule-of-thumb method for sizing up a specimen which resembles the Lilliputian way of measuring Gulliver for his coat. A Sperm Whale upon arrival at the surface will spout once for every minute that he has been down. If interrupted and driven from the surface, he will not stay down any protracted period until after he has completed his full quota. This is known as "having his spouts out." The average whale will stay down one minute for each foot of his length. He will spout that same number of times upon reaching the surface and will displace that same number of tons of water. In other words, a sixty-foot Sperm Whale weighing sixty tons will stay down while feeding sixty minutes, and will spout sixty times when he comes to the surface. This is relatively accurate. A whale of that size would probably stow down eighty or more barrels of oil.

When I went whaling I knew, as every other boy in New Bedford did, exactly what constituted a big whale, for I had been brought up in the tradition. A big Sperm Whale, for instance, was a whale that cut-in something over eighty barrels; anything over ninety barrels was a giant. But a whale that cut-in a mere forty-five barrels was just average, and required no comment in the workaday world. I could sniff at such a whale as well as the next man. I knew that the whale was the biggest animal that ever lived. Length, however, was a matter little discussed in the old seaport, for a barrel is the whaleman's yardstick, and the thickness of blubber and the condition of the case were the important features of any catch. If we had taken a whale as long as the ship, I shouldn't have batted an eye; that is, I shouldn't unless he cut-in over ninety barrels of oil.

The third whale I saw at close quarters was considerably smaller than the ship. I'd been in the boat that lanced him and I felt that he was a very satisfactory whale indeed. My eye told me that he was somewhat bigger than the eighty-five-barrel bull we'd taken just a week before, and I was glad of it, and was looking forward to my next whale. It wasn't until we had our pots full of case matter and the fires well started that I had my first misgivings about him. I overheard Mr. Smith say to Captain Higgins — "Most I ever saw." It was then I had an inkling that something had passed me. So I asked my first question — "Was that a big whale?" Captain Higgins looked a little surprised at my evident earnestness and said simply, "You'll never see a bigger."

When the case was drawn off, it measured exactly thirty-one barrels of spermaceti. Captain Higgins said this was five barrels greater than any he had ever heard of, and I have never been able to find a record of one so large, nor any captain who has seen one. Scammon says that "as much as fifteen barrels of oil has been obtained from the case of a Sperm Whale." W. M. Davis ["Nimrod of the Sea"] mentions a twenty-seven barrel case. This appears to be the biggest previously on record. Our whale stowed down a total of one hundred barrels of oil, but unfortunately he never was measured accurately.

Since a whale is the biggest animal that ever inhabited this earth, it seems to me that the limit of his size is one of the most interesting facts concerning him. For this reason I shall give some figures that are by no means representative — they are whalemen's records of the giants of the species.

Thomas Beale, who was the surgeon of an English whaler, records an eighty-four-foot Sperm Whale which he measured off the coast of Japan. Captain Thomas Sullivan, in 1866, on the Australian grounds, in the *James Arnold*, took a 137-barrel bull Sperm Whale ninety feet long. The flukes were eighteen feet wide, the case was twenty-two feet long, the jaw was eighteen feet long, and the forehead stood thirteen and one-half feet high. Captain William I. Shockley on the bark *Louisa* in 1875 off the West Coast of Africa took a 130-barrel Sperm Whale with a jaw nineteen feet long. The jaw of a bull Sperm Whale is between one fourth and one fifth of his length. In 1866 Captain Martin Malloy, of the bark *Osceola 3rd*, brought home to New Bedford a sperm whale's jaw measuring nineteen feet two inches. ["Fishery Industries of the United States," p. 262.] This was taken from a 115-barrel whale that stove three boats and then attacked the ship, biting away most of the cutwater. Unfortunately few whales were ever measured, for a whaleman scorns linear measurement. But if these whalemen's records are accurate, it would appear that the hundred-foot Sperm Whale is not an impossibility. These are big figures, but other whalemen have accepted them, and the whalemen that I have known were not prone to exaggerate their daily commonplaces. Captain William I. Shockley states that the number of barrels of oil taken is no correct gauge for the length of a whale, as the amount is largely dependent upon his condition. "If a Sperm bull has been with a lot of cows or has come recently from the Indian Ocean into the Atlantic in the breeding season, he will be poor, without much head oil; but if taken on the feeding grounds while still feeding he will be fat. A sixty-barrel Sperm may be either fat or thin, the average length would be from forty-five to fifty feet, the jaw eleven to twelve feet long. I should guess the age of such a whale to be from three to four years. I have helped to take between 650 and 700 whales and have hauled calves out of cows alongside plenty of times eight to twelve feet long."

Captain Shockley, when asked to estimate the length of his 130-barrel, nineteen-foot jaw whale, evaded a direct answer by writing me that "A sixty-five or sixty-eight-foot Sperm Whale is a good-sized one." Dr. Lucas estimates that Captain Shockley's whale was between seventy and seventy-eight feet long.

Scientists are apt to be skeptical about such figures as those given, but it should be borne in mind that there are giants of every species, and that no scientist of note, save Beale, a surgeon, not a naturalist, ever had a good opportunity to examine Sperm Whales until long after the herds were greatly depleted and the fishery on the wane.

The girth of a large Sperm Whale is not less than thirty-six or thirty-eight feet. The forehead slopes forward and the tip of the snout or skull is several feet back of the line of the spout-hole. The end of the jaw is recessed two or three feet under the snout. The length of the pectoral flipper is not much over five feet and it is probably used more for guidance than propulsion. His eye is little larger than an ox's. He cannot see directly backward or forward, having a blind sector in these directions equal to between forty and fifty degrees astern and about ten degrees ahead. At his ordinary gait the Sperm Whale can probably see the "loom" or shadow of any sizable object on the surface. When swimming rapidly, the head is lifted from the water and the tail depressed, and in this position he is able to see horizontally forward, as the underside of the head has "hollow-lines." The battering-ram-like forehead would be sufficient protection in almost any collision, and there is nothing alive in the sea that will not make way for him.

The Bowhead Whale sees much better ahead, having almost perfect vision in that direction, but he has an even wider blind sector astern. This provision is taken into

THE WHALE

account by whalemen in directing their attack. Since the Bowhead Whale's tail is very active, he can "sweep" his flukes "from eye to eye"; "going on" requires nice calculation. The usual procedure is to pull up the boat from the right rear quarter, just clearing the tip of the fluke. In this position the whale cannot see the boat, which carefully avoids his wake, as the drag of its weight in the eddy of the flukes would be instantly felt by the whale, causing him to sense danger and sound before the boat could get fast. The bow of the boat is brought forward to a point about opposite the flipper, keeping entirely out of the range of vision. The mate gives the order to dart, and the same instant, with a single sweep of his long steering oar, he swings the stern out at right angles to the whale (lays the boat on). The next instant the order "Stern all!" directs her out of danger. If the boat has "gone on" under sail, the iron may be "pitch poled" — that is to say, darted a considerable distance, in which case the boat is instantly put about, or kept off, and the crew with paddles assist her out of range of the flukes. In laying on, because of the great danger from the flukes, the boat is not bumped into the Bowhead Whale, "wood to blackskin." This latter method is confined to the Sperm Fishery.

The Sperm Whale is also approached from the rear and from the right side, if possible, since this gives a right handed harponeer a better stance, and lessens the danger of fouling the line on the second iron which rests in the boat crotch, a projection in the starboard gunwale.

As the Sperm Whale sees better astern than the Bowhead Whale, the boat generally pulls directly across the tip of the right fluke. When the boat is abreast the flipper, it is laid on with the men still pulling and the side of the whale acts as a backstop when the boat bumps into it. The eye of the Sperm Whale is a short distance above the angle of the mouth and the opening of the ear is a few inches in back of the eye. Although the ear is barely a quarter of an inch across, the hearing is very acute.

There are longer whales than the Sperm and Whalebone Whales, although none is bulkier. Of these, the Sulphur Bottom, or Great Rorqual (also called Blue Whale), is the longest. He is also a fast whale; a rate of fifty miles an hour in his initial sprint is frequently claimed. Murdoch ["Modern Whaling and Bear Hunting," p. 33] says he "has seen 2160 feet of line fast to a Sulphur Bottom run out in a few seconds at the rate of sixty miles per hour, with the engine going eight knots astern and brakes on." The accounts of his speed are probably exaggerated. Old whalemen say the Finback is the fastest whale. Racovitza says the sulphur bottom's spout is fifteen meters high (about fifty feet). Dr. R. C. Murphy states that specimens have been taken by the southern fishermen up to one hundred and eight feet in length. No successful way of taking these whales with whaleboats was ever discovered. They were too fast, for one thing, and moreover, they usually sank the instant they were killed.

The speed of the Sperm Whale is generally given as three or four miles an hour when undisturbed, and eight or ten when hurried. I have been fast to a whale in a moderate sea when two men had to bail constantly to keep out the water that came over the bows, and the ship, following us with a fair wind, was hull down in ten or fifteen minutes. W. M. Davis ("Nimrod of the Sea") states, "when struck he will frequently go twenty or twenty-five miles per hour for a short time." This to my mind is about correct. I have been in motor speed boats at better than forty-five miles per hour, and found it a tame performance after a "Nantucket Sleighride" behind a gallied Sperm Whale.

All whales upon sounding leave behind them a smooth oily area at the spot where they went down. The old English name for this spot was "glip," the American name is "slick." Whalemen have always maintained that, if a boat pulls into the slick or crosses the path of the whale between him and his slick, the whale is instantly conscious of it and as a result is gallied.

My own observation is that whalemen will go to a considerable amount of additional labor rather than cross the slick. As they are practical men, it would seem they must have strong convictions to justify this practice. But whalemen, like other fishermen, are strong believers in luck, and it is probable that they keep out of the slick for the same reason that most landsmen do not pass under ladders, and that the story of the whale's sentient connection with his slick is a whaleman's myth.

Many writers have expressed the opinion that the slick is an oily emanation from the skin of the whale. But there appear to be no glands to account for this. Racovitza says the slick is undoubtedly oil, which he traces to the excreta. Andrews scouts the idea of oil, and says the slick is "produced by suction and interrupted wave action." My own opinion is that the slick is mucus from the spout-hole, emitted at the conclusion of the final spout, before sounding, the purpose of which is to seal the tightly closed spout-hole. It is never observed in the wake of a whale except at the exact spot where he turned flukes.

The Sperm Whale is generally credited with the possession of some means of distant communication with other whales. When a whale is struck, other whales, miles away, but visible from the masthead, immediately express by their actions that they are aware that something untoward has occurred; they either take alarm, or, if they are cows, come to the assistance of the whale that is in trouble. It seems probable that the commotion made by the whale when struck, due to the excellent acoustics of the water, is heard and understood by his distant companions. Cow whales will stand by one another when attacked and a cow will not desert her calf even after it is killed. But bull whales have no sense of loyalty; unless in surly mood they turn flukes when a comrade is in trouble.

When feeding or at play, the whale has a practice which is termed "lobtailing."

Standing almost on his nose, with his tail vertically out of the water some thirty feet or more, he waves it violently, hitting the water at each stroke with a sound that can be heard for miles. The water is a mass of seething suds, and the air is charged with spray. But this practice of "lobtailing" does not disturb his comrades in the least, no matter what their proximity. So it is evident that mere commotion is not the cause for their alarm when another whale is struck by a harponeer.

Most varieties of whales have the habit of "breaching." The Sperm Whale, when he leaps clear of the water, does not make a finished "porpoise dive." His practice is to shoot out headforemost and then to drop flat on his belly or side with a resounding whack. For this reason, it has been assumed that he breaches to free himself of parasites. He seems, however, to do it quite playfully. I have seen an eighty-five-barrel Sperm bull leap clear of the water, so that the afternoon sun was framed for an instant under his hurtling form.

When a Sperm Whale first comes to the surface, the hump is the only part of his anatomy out of water. His breathing and his swimming are extremely rhythmical. The hump settles as the distended spout-hole rises, and the spout commences at the instant the hole is clear of the water. The whole upper bulk of the slowly lifting head emerges clear of the water several feet, then "pitches" in continuation of the same movement, as the hump reappears. His ordinary swimming consists of a steady up-and-down churning of the flukes, but when swimming slowly there is also apparent a slight side-to-side waggle of the flukes similar to the motion of a scull oar; on one stroke the tail approaches the boat, the following stroke it moves away. The movement is more pronounced in the swimming of the Bowhead. At an ordinary pace the stroke of the flukes is neither rapid nor extensive, but when the Sperm Whale is gallied, the back rounds out sufficiently on the downstroke to show that the body at the hump is bent almost at right angles. The head on the upstroke rises six or eight feet from the water. The flukes, of course, have to take a reverse angle in order to propel — that is to say, on the downstroke they are tilted upward, and on the upstroke they are tilted downward. In both strokes, then, his profile, if visible, would present a rather abruptly reversed curve. When he goes at top speed his head lifts entirely from the water until the jaw is in view, and the head rises and pitches with the rapid beat of his flukes, but does not disappear beneath the water.

The flukes of the Sperm Whale have little sidewise play, but his jaw is exceedingly mobile. In attack he frequently rolls from side to side with his head out of water, and in this position the scope of his jaw is tremendous, being equal to twice its length. A boat fifteen feet away at either side is in imminent danger from a rolling Sperm Whale's jaw.

A Sperm Whale will sometimes stand upright, his whole head out of the water, and, with his small pig-like eyes above the surface, he will then "mill" slowly around in a complete circle in order to view the boats, bobbing up and down as he turns. He does this when first alarmed and it is not considered a favorable attitude by the whalemen. This action is termed "pitch-poling" from its fancied resemblance to the preliminary gestures of the harponeer before darting his iron.

When attacked the Sperm Whale may stand and fight instead of running, as other whales do. His most dangerous offensive position is "jawing back." Rolling over on his back, with head out of the water, he lashes out with his jaw, snapping and chewing to splinters everything within reach.

The barks *Commodore Morris* and *Atlantic*, while gamming in the North Atlantic in 1879, had seven boats chewed up by one whale — they succeeded in taking him the following day.

The Sperm Whale is a born fighter. Although in combat he generally directed his efforts against the boats which attacked him, on the few occasions when his mind gathered that a ship also was in some way implicated, the results were both startling and disastrous. A number of whalers have limped into port so damaged that they had to "lay up" for repairs, and on several occasions a ship has actually been sent to the bottom. The most recent of these was the bark *Kathleen*, in 1902. The whale charged the ship only once, but she sank five minutes after the last man scrambled out of her. One of the boats containing nine men made Barbados, a distance of 1060 miles, in nine days, subsisting in that period on five gallons of water and the few biscuit that were in the boat's lantern-keg. The other boats were picked up by passing ships. The *Ann Alexander* of New Bedford and the *Essex* of Nantucket were both rammed and sunk by enraged bull whales that had previously been attacked by the boats. It speaks well for the order of a Sperm Whale's intelligence that in his deliberate assault upon a ship he generally discarded his ordinary weapon, his fighting jaw, which would have been useless against so huge and armored an adversary, and resorted to ramming with his massive forehead, the only means at his command that could possibly prevail.

Sperm-oil is still used as a lubricant for certain kinds of machinery and also as an illuminant in railroad lanterns, as it is not easily extinguishable. Spermaceti is required in paraffine candles to prevent their wilting, and is also used as a base in high-grade cosmetics. The occasional Sperm Whale taken by the shore fisheries is sufficient to supply these demands.

Whalebone, which had a hundred uses in its day, is now scarcely required at all. Once it was in demand for whips, stays, and haircloth. But every one of these commodities, for which it was once essential, has by a curious quirk of circumstance gone entirely out of fashion. At one time whalebone brought upward of six dollars a pound at wholesale, but to-day it does not possess even a curio value.

In 1908, Dr. Frederic A. Lucas ("The Passing of the Sperm Whale," New York,

1908) wrote: "Nothing can possibly prevent the extermination of the bowhead but the discovery of some perfect substitute for whalebone, and there seems not the slightest probability that this will be done." But the perfect substitute was found, made from the quills of feathers, and for a time served its purpose. To-day the substitute is as obsolete as the original.

Ambergris has always been the most valued product or by-product of the whale-fishery. It has the rare property of intensifying and suspending any odor to which it is added, and for this reason it is invaluable in perfumery. At present this appears to be its sole commercial use. Formerly it was supposed to have a medicinal value, and was also used for incense and as an aphrodisiac. Its exact nature is not defined, although it is known to be a foreign substance which forms in the alimentary canal of a Sperm Whale, and when found generally has the beaks of squid embedded in it. Presumably it is of a scabious nature, and results from a squid's beak having punctured the wall of the intestine. Nature forms a cicatrix about this beak in an effort to throw it off. Eventually this either passes naturally out of the whale's system, or it clogs up the passage and the whale dies. Ambergris is found only in "sick" whales, and a thin whale is always carefully searched by whalemen for this precious material, which is worth more than its weight in gold. Clean ambergris has very little odor; what there is of it is pleasant. Its value is so great — sometimes as much as several hundred dollars a pound, although the price varies greatly — that it is used only in combination with the most valuable of scents. It is most frequently employed with attar of roses.

These were the products of the old whalefishery, the rest of the whale was cast adrift.

When modern steam whaling came into vogue, a rendering station converted every vestige of the carcass into marketable produce — the flesh into prime canned beef for the Orient, the bones and refuse into fertilizer, the flippers and flukes into glue. The oil of the fin whale produces a considerable amount of commercial glycerine. Whale-meat-meal is made from the flesh, and is fed to cattle. New methods are far more provident than the old. But Sperm and Bowhead Whales are not taken.

The final issue of the "Whaleman's Shipping List," published December 29, 1914, quoted sperm-oil at $.48, and in the same issue, speaking of bone, said: "We are unable to quote any sales — there does not seem to be any demand for the large stock on hand."

For ten years no New Bedford whaler has come out of the Arctic. So far as I know, not a pound of whalebone is now held for sale commercially in America. In the meanwhile, Bowhead and Sperm Whales are increasing, and as a result of this increase, with inferior crews and officers, and with poorly found vessels, the last scattering voyages out of New Bedford were uniformly successful. A full voyage took half the time required twenty years ago — a quarter of the time it took sixty years ago. Consequently, a recent voyage with sperm-oil at less than fifty cents a gallon "broke even," where "one dollar oil" in the fifties frequently meant a "losing voyage." And this was despite the increased cost of outfitting.

So, at a time when the steam fishery is seriously threatening the extermination of the Humpbacks, the Finbacks, and the Sulphur-Bottoms; the Bowhead Whales of the Arctic are increasing and Sperm Whales are becoming so numerous that eventually they may prove an actual menace to navigation. The Sperm Whale will never realize that he isn't the biggest thing afloat. So it is not at all improbable that in the near future it will become necessary to put a bounty on him. He is hard meated, and the fact that soft meated Finbacks have been run down with impunity is no guarantee that any ship is safe to strike a one-hundred-ton sperm bull while running at full speed.

STOVE BOAT

A fighting sperm bull, entangled in whale-line, is chewing up a boat. This gives an idea of the relative size of the whale's under jaw, the flatness of his forehead, and the shape and position of his spout-hole

In the collection of the Nantucket Whaling Museum

THE LAST WHALER
The Charles W. Morgan tied up at New Bedford

FITTING OUT THE SUNBEAM, 1904
In the collection of the New Bedford Public Library

FITTING OUT BARK ANDREW HICKS
Painting owned by Mrs. John W. Knowles

THE GREENLAND FISHERY
Sketch owned by Dr. F. A. Lucas

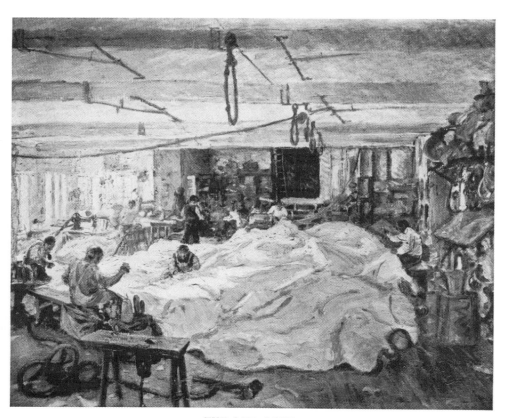

THE SAIL LOFT
A squaresail laid out in the Briggs and Beckman loft
In the collection of the Mariners' Museum, Newport News

CAULKING A WHALER

"THE FOOT OF THE STREET"

Centre Street with whaling warehouses at either side and the Morgan fitting out at the wharf
Painting owned by Mr. George H. Taber

BARK SUNBEAM BECALMED
Painting owned by Mr. R. R. M. Carpenter

BOTTLE-NOSE PORPOISES
Painting owned by Mr. Walter G. Carpenter, Jr.

THE DAY BEFORE SAILING
Painting owned by Mrs. Eben E. Whitman

THE WANDERER TIED UP
In the collection of the New Bedford Whaling Museum

GATHERING FOG
Painting owned by Mr. William F. Sellers

THE GREYHOUND OUTFITTING
Painting owned by Mr. A. Felix du Pont

OUTFITTING THE CHARLES W. MORGAN, 1916
This shows the earliest type of boat-davit
In the collection of Wilmington Society of Fine Arts

THE BARK COMMODORE MORRIS OFF DUMPLING LIGHT
Sketch for an overmantel in the home of Mr. A. A. Houghton

THE STERN OF THE WANDERER
Painting owned by Mrs. A. R. Gould

A WHALER AT UNION WHARF
Painting owned by Mr. Arnold C. Gardner

THE WANDERER, FROM FISH ISLAND
Painting owned by Mr. Frederic A. Delano

CHAPTER IX

GEAR AND CRAFT

THE whale harpoon used up to the middle of the nineteenth century in the American Fishery did not differ essentially from the iron used in the early days of Spitsbergen. Van Oelen, in his book on whaling ("De seldsaame en noit gehoorde Walvischvangst," J. A. Van Oelen, Leyden, 1684), shows a stereotype two-flued iron, and an illustration in Thomas Edge's account of his voyage to Spitsbergen in 1611 ("Churchill's Voyages," London, 1744) shows the same variety. The earliest irons and lances had a shoulder and spike at the base, instead of the familiar socket. An eye was "grafted" on the shank at the socket or shoulder, and the whale-line was bent to this. Several of this type, left over from the early days of our own fishery, have been preserved in New England.

The English in the middle of the eighteenth century introduced return barbs, or beards; also called "stop-withers." These were short reversed barbs added to the inside of the flues. These irons were never used by the Americans, but the English considered them an improvement and persisted in their use long after the Americans had given up the common two-flued iron altogether in favor of the toggle.

Some time in the 1840's, the single-flued iron was tried out. The single flue was longer than the flues of the older iron, and it was supposed that the shank would bend slightly under tension, causing the single flue to turn more or less at right angles to the pull, giving it a greater bearing and consequently better holding properties. It is possible that this iron would have superseded the two-flued iron, since it immediately became popular. But with the entrance of American ships into the

Kodiak Grounds in 1835, American whalemen became acquainted with the walrus and seal spears of the Eskimos and Indians of the Northwest. This spear had a detachable head which pulled off the shaft after being darted, and turned at right angles to the pull of the line. The principle of this spear-head made a convincing appeal to the whalemen, and upon their return home with specimens, practically every whalecraft-maker in New Bedford turned to the problem of embodying this principle in an improved whale-iron. The Eskimos of Greenland used similar spears and there is evidence of much earlier attempts to fashion a whale-iron that would "toggle." Over one hundred harpoon patent applications were filed in Washington within the next few years, some with removable heads, some with pivoted or "toggle" heads, and some with hinged barbs. A number of these harpoons promised well, but one in particular, invented by Lewis Temple, a negro whalecraft-maker of New Bedford, in 1848, was of such extremely simple construction, and at the same time was so practical, that it was at once adopted to the exclusion of all others. With but one slight change, to permit easier manufacture, it has continued to be the standard harpoon of the fishery. The Temple iron offered less resistance in entering the whale, and, having toggled or turned at right angles under strain, it presented greater resistance against withdrawal than any other harpoon. To hold the head parallel with the shank until it had been driven through the blubber, there was a short wooden pin the size of a match. This pin broke the instant pull was exerted against the flue.

The modern harpoon or iron was mounted on a six-foot pole of oak or hickory with the bark still on. A round turn of the whaleline was tightly spliced about the shank of the iron where it joined the socket. The line was stretched and "stopped" twice to the pole, and an eye-splice was put in the other end, at a point about halfway up the pole. To this eye the whaleline was bent.

The account books of James Durfee, an early whalecraft-maker of New Bedford, show that while whaling was at its height his shop made 58,517 harpoons in forty years. There were between fifteen and twenty whalecraft shops in New Bedford and Fairhaven at this period. These figures, concerning but a single item of a whaler's equipment, give some idea of the amount of material required to keep the fleet at sea.

The old two-flued and single-flued harpoons were made of iron and were rough-sharpened, often with a file. But the lance-head and the Temple toggle-head were made of steel, the lance because it was used over and over again and had to hold its edge, the toggle for added strength, since the head was of cast metal and all the strain came upon the hinge-pin and its bearings.

It is safe to say that the Temple toggle was the most important single invention in the whole history of whaling, since it resulted in the capture of a far greater proportion of the whales that were struck than had before been possible. Since none of the other toggles ever came into practical use, it is not necessary to discuss them at length. A number of experimental irons are shown among the illustrations. Temple's toggle, at first called "Temple's Gig" was also frequently referred to as a "porpoise iron." It is probable that whalemen practiced with it on porpoises before they were willing to risk it on whales. The value of a single whale is so great that whalemen in the excitement of the chase would seldom resort to experiment, no matter what was planned in advance. Experimental irons were taken to sea in many ships, but few were ever darted.

The purpose of the harpoon is not to kill whales, although on rare occasions death may result. A harpoon attaches the boat to the whale, and the whale is then lanced to death when opportunity presents. The whale-lance has a petal-shaped blade sharpened on both sides to a razor edge. It has a shank five or six feet in length. The longer is for the Arctic Fishery, where whale blubber is thicker. The socket of the lance fits on the sharpened end of a six-foot pole, making an instrument twelve feet long. With this weapon the whaleman seeks to reach the "life" of a whale, as he terms a vital spot. The "life" is his lungs rather than his heart.

While Yankee whalemen still sailed as boat-headers, the fluke-spade was also a weapon of offense. To use this implement effectively called for the greatest skill and judgment, and also for courage and a steady nerve. Its purpose was to stop a running whale before he could reach the ice. The method was to pull over the churning flukes, and with a single thrust to sever the tendons at the small, completely hamstringing the whale. If there was the least miscalculation, up came flukes and a stove boat resulted. It is possible that this master-stroke has not been employed for thirty or even forty years, yet it was once the particular pride of a crack whaleman to be able to do the trick with precision and dispatch. The advent of explosives marked the end of this nice but dangerous practice.

The first successful bomb-lance (an explosive bomb fired from a heavy shoulder gun) was invented in 1852 by C. C. Brand, of Norwich, Connecticut. Another shoulder gun and bomb-lance, perfected by Eben Pierce, of New Bedford, a few years later, became the standard of the Fishery. The bomb or lance was a brass cylinder about fourteen inches long, with a pointed iron head and rubber or metal feathers. It had a time fuse, and exploded shortly after entering the whale. The whaleman could keep well out of range of the flukes, and the bomb hastened the killing very materially. Many whales were saved that otherwise might have escaped into the ice. But the gun was not so generally used in the Sperm Fishery as there is no ice in Sperm whaling grounds to offer cover. Time was of little importance, and bombs represented a certain expense; slight, it is true, nevertheless they were not to be wasted. But the greatest objection to explosives in the Sperm industry was the fact that Sperm Whales generally travel in pods or schools, and if care were taken to make no unnecessary noise to disturb and gally them, there was always a chance that several boats would take whales from a single pod. The *Sunbeam* one day shortly after I had left her had eleven whales alongside as the result of one lowering. While I was still aboard, a bomb-lance was fired just once, and that time the breech carried away and injured the third mate's thumb.

A "darting-gun," subsequently invented by Eben Pierce and Patrick Cunningham, of New Bedford, proved a more practical weapon than the shoulder gun, and was used almost to the exclusion of other whalecraft in the last days of the Bowhead Fishery. This darting-gun was stockless and was mounted on the end of a harpoon-pole. A harpoon was loosely slotted to the side of the gun-barrel. The whole instrument was darted in exactly the same manner as a harpoon. When the iron had penetrated to the hitches, a trigger brought up against the whale and the gun went off, shooting a bomb into the animal. The gun was then recovered by a lanyard attached to the pole, the iron remaining in the whale. The whaleline was spliced to a ring in the shank end of the harpoon. Frequently, the first dart both fastened and killed the whale. The darting-gun was more often used in the Sperm Whale Fishery than the shoulder gun. It proved an expeditious means for taking a lone bull, but, like the bomb-gun, it was sparingly used when whales were herded.

It may have been an indication of the decadence of the industry, or of the increase and consequent easier capture of whales, or it may even have been because of the need of economy; but certain it is that in the last twenty years of the whaling industry, a gradual return was made to the old toggle-iron and hand-lance. Explosives, although still carried, were infrequently resorted to. But again, this may have indicated merely that the Portuguese, into whose hands the industry was drifting, did not have the same assurance with explosives that Yankees possessed.

In the British Fishery, a swivel harpoon gun was tried as early as 1730, and with fair success, according to Scoresby. But about this time the British Fishery had one of its periodical slumps, and by the time it recovered, the gun had been lost sight of. The method was revived in 1772, and thereafter for a period of forty or more years the "Society of Arts" paid a premium not only to any one who would add an improvement to the gun, but also to any one who was successful in taking a whale by means of it. (Scoresby, p. 79.)

This gun pivoted at the bow of the boat, and hurled a harpoon with line attached. Although its range was greater than the hand-iron, about one hundred feet, it was much less accurate, since it could not accommodate itself to the roll of the boat. For that reason it was never favored among American whalemen, who on occasion have darted the hand-iron successfully over thirty feet. In the fifties, the Greener bow-gun was successfully used in the comparatively smooth waters of the "California Bay Fishery" in "Humpbacking" and "Devil-fishing."

GEAR AND CRAFT

In 1833 the sloop *Fame* of Nantucket sailed "in search of whales, sea serpents, etc.," and was armed with a harpoon loaded with prussic acid. A French ship similarly equipped having reported a number of her men killed from having poison from the whale's carcass enter abrasions and cuts in their skins, the Nantucketers took alarm, and, it is said, did not attempt to employ their new weapon.

The last important invention in whalecraft was made by the Norwegian Svend Foyn, in 1860. His invention consisted of a cannon firing an explosive harpoon weighing over one hundred pounds which had four hinged barbs on its head. This was carried at the bow of a small steamer, and a four-and-a-half-inch hemp cable was used for a whaleline. With this invention it became possible to take Sulphur-Bottoms and Finbacks, neither of which had ever before been successfully hunted, and it was also employed to take Humpbacks.

These whales usually sink when killed, but the cable and a donkey-engine haul them to the surface where they are inflated with compressed air to keep them afloat. The ships always operate in connection with a shore rendering plant, or else a large mother ship, which serves the same purpose.

A description of the Basque Fishery, by Ambroise Paré in 1564, tells that after the whale hunt each harponeer was rewarded according to the number of his harpoons found in the body of the whale. This shows that the harpoon of the period was closely related to the arrow, being used merely to wound the whale and not to hold him in check. Apparently, the whaleline was not in use at that time.

Purchas his Pilgrimage, London, 1617, p. 923, describing the Greenland Fishery, states: "When they espy him on the top of the water — they row toward him in a Shallop, in which the Harponier stands ready with both his hands to dart his Harping Iron to which is fastened a line of such length that the Whale which suddenly feeling himself hurt sinketh to the bottom may carry it down with him — the Harping Iron principally serving to fasten him to the Shallop."

The New England settlers, however, were not familiar with the methods of the Greenland Fishery, and they patterned their attack after the Indian method of taking whales. Their first harpoon, which bore the usual arrow-shaped head, had no line leading to the boat, but was attached by a "short warp" to a square piece of wood called a "drug" which was thrown overboard as a sort of sea-anchor for the whale. As many of these irons and drugs were "fastened" as opportunity allowed. The cumulative result of their pull was to slow down his flight, and eventually exhaust and stop him, so that he could be killed.

Waymouth's "Journal of his Voyage to America" in 1605 (published in the Massachusetts Historical Society Collections, Series III, vol. 8) interestingly describes the Indian method:

One especiall thing is their method of killing the Whale, which they call Pow-dawe; and will describe his form; how he bloweth up the Water; and that he is twelve Fathoms long; and that they go in company of their King with a Multitude of their Boats; and strike him with a Bone, made in fashion of a Harping Iron fastened to a Rope, which they make great and strong of the Bark of Trees, which they veer out after him; then all their Boats come about him as he riseth above Water, with their Arrows they shoot him to death; when they have killed him and dragged him to Shore, they call all their Chief Lords together, and sing a Song of Joy; and those Chief Lords whom they call Sagamores, divide the Spoil, and give to every Man a share, which pieces so distributed, they hang up about their Houses for Provisions; and when they boil them they blow off the Fat and put to their Pease, Maize and other Pulse which they eat.

The Greenland Eskimos' method of capture, which is described in "Harris's Voyages" (London, 1748, vol. II, p. 382), is also suggestive of our early shore methods, and in one detail, the use of the spear or lance, was an advance over the Indian way:

They [the Greenland Eskimos] surpass most other Nations, for their way of taking Whales and other Sea Animals is by far the most skilful and easy. When they go whale catching, they put on their best apparel, as if they were going to a Wedding Feast, the Gronelanders fancying that if they did not come cleanly and neatly dressed, the whale would shun and fly from them. The manner of their expedition is thus; about fifty Persons, men and women, set out together in one of the large Boats called a Cone Boat. The Women carry with them their Sewing Tackle, consisting of Needles and Thread, to Sew and Mend their Husbands' Spring Coats or Jackets, if they should be torn or pierced through, as also to mend the Boat in Case it should receive any Damage. The Men go in search of the Whale, and when they have found him, they strike him with their Harpoons, to which are fastened lines or straps, two or three fathoms Long, and made of Seal Skins, at the end of which they tie a bag of a Whole Seal Skin, filled with Air like a Bladder, that the Whale, when he finds himself wounded, and runs away with the harpoon, may the sooner be tired, the Air Bag hindering him from keeping long under the water. When he grows tired and loses strength, they attack him with Spears and Lances till he is killed, and then put on their Spring Coats, made of Dressed Seal Skins, all of one Piece, with Boots, Gloves and Caps, laced so tight together that no water can penetrate them. In this Garb they jump into the Sea, and begin to slice the Fat of him all around the Body, even under the Water; for in these coats they cannot sink, because they are always full of Air; so that they can, like the Seal, stand upright in the Sea, Nay, they are sometimes so Daring, that they will get upon the Whale's Back while there is yet Life in him, to make an End of him, and cut away his Fat the sooner.

The inflated sealskin bag referred to, served the same purpose as the wooden "drug" and was later adopted by American whalemen after they had begun to whale in northern waters. Either a "drug," or a seal or blackfish "poke," as it was called,

GEAR AND CRAFT

became part of the regular equipment of every modern American whaleboat. In modern times it was tied on only as a last resort when it was evident that the whale was about to "take the line." When the bitter end was in sight, the poke was blown up by mouth and tied to the whaleline outside the chocks with a rolling hitch, before being tossed overboard.

William Douglass ("Summary, Historical and Political, of the British Settlements in North America," London, 1760, vol. II, pp. 296–98) describes the "drudge or stop-water" as "a plank of about two feet square with a stick through its center; to the further end of this stick is fastened a tow-rope called the drudge rope, of about fifteen fathom; they lance, after having fastened her by the harpoon, till dead." Douglass further says that the harpoon line or "fast" is "a rope of about twenty-five fathom."

There has been much discussion about the use of the "drug" and it has been very generally contended that the Americans never adopted the Indian method of fishing.

Scammon (p. 204), states, "We are of the opinion, however, that the Colonial whalers did not follow the Indian mode of whalefishing, for it is well known that the British whalers, as early as 1670, used the line attached to the boat, and so far as the drugs or 'drogues' are concerned, they are used at the present day in cases of emergency."

Other lesser authorities support this statement, but A. Howard Clark ("Fishery Industries of the United States," Washington, 1887, p. 49) finds the drug still in common use in shore whaling in North Carolina in 1880. "As soon as the whale is harpooned, the 'drug' is thrown over, and when he turns to fight, the fishermen, armed with guns, shoot him with explosive cartridges." We see that the drug here is part of the initial offensive, and not a final resort as contended by Scammon. Clark further states that "The whale fisheries of Beaufort seem to have been prosecuted continuously for a long period of years, and the oldest inhabitants are unable to give any information of their origin."

The Honorable Paul Dudley, a colonial Chief Justice of Massachusetts ("An Essay upon the Natural History of Whales," vol. III, Transactions, Philosophical Society of London, 1725), gives the following:

Our People formerly used to kill the whale near the shore but now they go off to sea in sloops and whaleboats. Sometimes the whale is killed by a single stroke, and yet at other times she will hold the whalemen in play near half a day together, with their lances; and sometimes they will get away after they have been lanced and spouted thick blood, with irons in them, and drogues fastened to them, which are thick boards about fourteen inches square.

This shows that both methods were in practice in 1724 in the Right Whale Fishery.

About this time the Sperm Whale Fishery was initiated, and it proved to be an entirely different sort of game. The Sperm Whale was a larger and more ferocious animal than the Right Whale, and whalemen the world over had always avoided him. It is probable that the towline was not used even experimentally for a considerable period of time after the institution of this fishery.

J. Ross Browne ("Etchings of a Whaling Cruise," New York, 1846) states that an American whaleman informed him that his great-grandfather had been employed in a Sperm whaling expedition off the American coast which had been quite unsuccessful. Many whales had been struck, but had escaped, carrying away both drugs and lines. The captain proposed "fastening" to a Sperm Whale, explaining to his crew how he planned to conduct the boat under the unusual circumstance. The men were terrified, and in the night dumped all the whaleline overboard in order to divert the captain from his monstrous plan.

The log of the sloop *Manufactor* of Dartmouth which sailed on a Sperm whaling cruise in 1756, quoted elsewhere in full, contains the following:

3rd Day, ye 20th, was fresh N.N.W. Winds We saw plenty Sparmecitys and Struck four But Could not get one Lat. $37°.-10^m$.

4th Day, the 21st. Was moderate S.W. Winds. Saw many whales and I struck one and got in a 2nd and a *tow iron* after which Z. Gardner Lanc'd her and She Stove and over-set our boat but did not hurt a man. Afterwards we got her But lost her head.

It will be noted that on the 20th they struck four whales and got none, evidence of either poor method or poor handling; that on the 21st they struck one and "got in a 2nd and a tow iron." The inference is clear that the first iron was *not* a tow-iron, since the tow-iron is specifically mentioned in contradistinction in the same sentence, and also that the four irons used the previous day were not tow-irons, for the same reason, and also because not one of the four whales was taken, an inconceivably poor average if a fast line had been employed. It will be further noted that untoward things began to happen the instant the boat was really fast to the whale with a tow-iron.

That the short warp and drug were still commonly used in attack in 1761 is shown by quotations from the log of the *Betsey* of Dartmouth (Daniel Ricketson's "History of New Bedford," pp. 62–64):

Aug. 22, 1761. This morning saw a spermaceti and killed her. Saw a sail to leeward, standing westward.

Sept. 3. This morning at eight saw a spermaceti; got into her two short warps and the tow iron; she drawed the short warps and the tow-iron, and ran away. In the afternoon came across her; got another iron in, but she went away. Judge ourselves to be nigh the Banks.

Sept. 6. Yesterday afternoon saw whales; struck one, but never saw her again.

GEAR AND CRAFT

A record of encounter is given with seven whales, and the tow-iron is mentioned but once; and this occasion is the only one where it is mentioned that the iron "drawed." In the entry of September 6th, it states merely that the whale was "never seen again" after being struck. If a towline was used when the whale succeeded in getting away, one of three things must have occurred: either the iron "drew," the line parted, or the whale "took the line." Any one of these three circumstances would be of supreme moment to the whalemen and they would not have failed to record it in their log.

For these reasons it may confidently be asserted that even so late as 1761 it was not yet the common practice of American whalemen to "fasten" to Sperm Whales, although they were experimenting with the method. The fact that not a single Nantucketer had been killed in the whalefishery previous to 1760 (Macy, "History of Nantucket," p. 31) is contributory evidence of this fact, since death in the Sperm Fishery shortly after, when we know that a fast line was being used, was a very common occurrence.

At the time the drug was employed, Sperm Whales populated the near-by waters in immense herds. The early whaling sloops generally manned a single boat; infrequently two. By practicing this method of fishing, one boat often was able to take several whales from a single pod. The record of the log-books shows that even when a whale got away there was a fair chance of encountering him later, or of picking up a carcass killed by some other whaler. It was not until whales grew scarce that the method became impractical. "Fastening" to a whale was a more dangerous method, but fewer whales escaped. As soon as whalers began to lower several boats simultaneously, "fastening" became the rule.

St. John in 1782 ("Letters from an American Farmer," London, 1782) describes the line coiled in the middle of the boat, with the end hitched to the bottom. This definitely places the date, when "fastening" became the rule, as somewhere between 1761 and 1782.

The practice of tying the end of the line to the boat appears to be a British one. Scoresby describes it in 1820 (p. 277). The British boats were much heavier and sturdier than the American, and if one was hauled under she was strong enough to suffer no damage. The boat acted as an effectual drug, so that the whale was usually taken by another boat after the crew had first been picked out of the water. Scoresby calls this "giving the whale the boat." In their warm sealskin clothes with oars to float them, the men were in little danger, and the whale was of far greater value than the lost boat-gear.

The Yankees introduced the custom of not hitching the whaleline. The American boat was lighter and consequently more fragile than the British boat, and her bows were much sharper, so that she pulled under more easily. Sperm whaling took place always in the open sea and boats generally became widely separated. In case of accident, it was well to have the wreck of the boat to cling to. Moreover, the warm waters of the Sperm whaling grounds were shark-infested. It has often been recorded that American boats never tied their lines, and I am inclined to believe that St. John, who was not a whaleman, although an excellent historian, is mistaken on this technical point.

The early whaleline was of hemp. According to David Steel ("Elements and Practice of Rigging and Seamanship," London, 1794), the circumference was two and a fourth inches. Scoresby mentions the same size in 1820, and Luce in 1860. The whaleline made by the Plymouth Cordage Company and the New Bedford Cordage Company, which has been the standard line for many years, is about three fourths of an inch through, modern rope being measured by diameter. It is made of the longest obtainable manila fiber, "slack" or "long-laid," the fibers being spun while moistened instead of being oiled, as in ordinary rope. It is thirty-nine-thread stuff, thirteen threads to the strand, and is extremely pliable even when new. Its tensile strength is about fifty per cent greater than ordinary rope of the same size, which places the breaking point of whaleline at about three tons deadweight. In the steam whale-fishery, Italian hemp is used, from four and one half to six inches in circumference. Steel cable has been tried in this fishery, but is liable to kink. Weight for weight, manila whaleline is stronger than steel cable of equal length, although, of course, it is much bulkier.

In the modern whaleboat about twenty-five feet of the forward end of the line was coiled in the bow box (the depression forward of the clumsy cleat). This was called the box-line, or the box-warp, and was tied to the eye of the "first iron" with a double becket hitch. A "short warp" was hitched to the second iron, and a bowline in the after end of it encircled the main warp. The second iron was darted, if possible, but, if there was no opportunity to reach the whale, it had to be tossed overboard. After the whale ceased to take out line, the second iron, if loose, generally drifted back to the boat, and was recovered.

There were several different types of cutting-spades for various purposes. A bone spade had a long round shank. A head spade a long flat shank. A half-round spade, called a "gouge-spade," was used in cutting holes for blubber-hooks and chains. Deck-spades had wider cutting edges and shorter poles than those used on the cutting-stage. The boat-spade was narrow and light, with a pole not over six or seven feet long. It was required to make a hole in the head for a towline after a whale was captured, and also was formerly used in combat. Sometimes a whale was towed tail first, but this was poor practice, as a whale tows much more easily head first. If towed this way the tips of the flukes were cut off and a hole made for the tow rope.

Other bits of whalecraft are elsewhere described or illustrated. The boarding-

GEAR AND CRAFT 95

knife was used at the gangway to separate the incoming blanket-pieces; the mincing knife sliced the horse-pieces into "books" ready for boiling. There were blubber-gaffs for hauling blubber about, blubber-pikes for shoving and tossing it, and blubber-forks for pitching it into the try-pots. Skimmers and bailers were not whalecraft, since they were of copper and were made by a sheet-metal worker. The whalecraft-maker was an iron smith whose forge to the casual eye differed from the usual smithy only in the one detail of the anvil's horn being turned away from the fire, since it was not required for horseshoes or other curved objects.

Churchill's description of the Spitsbergen Fishery ("Churchill's Voyages," London, 1744) gives with considerable particularity the early manner of "saving" the whale:

When the whale is killed, he is towed to the shipps by twoe or three shallops made fast to one another. The whale is cut up as hee lyes floting crosse the sterne of a shipp the blubber is cut from the flesh by peeces 3 or 4 foote long, and being rased is rowed on shore toward the coppers. The peeces of blubber are towed to the shore side by a shallop and drawne on shore by a crane or carried by twoe menn on a barrowe to ye twoe cutters, who cutts them the breadth of a trencher and very thine, and by twoe boys are carried with handhooks to ye choppers. They place twoe or three coppers on a row and ye chopping boat on the one side and the collinge boat on the other side to receive ye oyle of ye coppers, the chopt blubber being boyled is taken out of the coppers, and put in wiker baskets or barowes throw which the oyle is dreaned and runs into ye cooler which is one half full of water out of which it is conveyed by troughs into buts or hogsheads.

Contemporary plates in the last part of the seventeenth century show the whale being cut-in at the side of the ship instead of across the stern. This is in agreement with an account of the Greenland Fishery in "Harris's Voyages" (vol. II, p. 388, London, 1749):

They haul him in close to the shipside and with great knives slice his sides, raising his blubber by a hook and a pulley which they lift up as they cut. Many of these great flakes they string upon a rope and so drag them on shore, where they are heaved up by a crane.

An engraving published about the same time in "Churchill's Voyages," illustrating Monck's "Account of a Most Dangerous Voyage to Greenland" (London, 1744), shows an entirely different method. The engraving pictures a whale on shore, alongside two great windlasses of the same sort that were formerly used on New Bedford wharves to heave down whaleships for repairs. The caption on the plate reads, "A Whale Female and the Windlais [sic] whereby the Whales are brought on shore."

This latter was the method followed by the Nantucketers in their early shore fishery, according to Obed Macy ("History of Nantucket," 1835), who says:

The process called Saving the whales after they had been killed and towed ashore, was to use a crab, an instrument similar to a capstan, to heave and turn the blubber off as fast as it was cut. The blubber was then put into their carts and carried to their try-houses, which in that early period, were placed near their dwelling houses, where the oil was boiled out and fitted for the market.

Some citizens of Eastham, Cape Cod, addressed the following petition to the General Court in 1706 ("Fishery Industries of the United States," Washington, 1887, p. 27):

All or most of us are concerned in fitting out Boats to catch and take Whales when ye Season of ye year serves... when we have taken any whale or whales, our Custom is to Cutt them up and take away ye Fat and ye Bone of such Whales as are brought in, and afterwards to let ye rest of ye Body of ye Lean of whales Lye on Shoar in lowe water to be washt away by ye sea, being of noe value nor worth Anything to us....

But when the Nantucketers began to fish off-shore, they cut-in their whales alongside, standing on the carcass and cutting with axes and spades. They stowed their blubber in casks, and brought it ashore to be tried-out. So late as 1853 the schooner *Armida* of Greenport sailed without try-works, making short voyages and bringing her blubber home.

The first record of try-works on shipboard is in the log-book of the ship *Betsey* of Dartmouth in 1762 (Daniel Ricketson, "History of New Bedford," 1858, p. 44). Under the date of September 3d is the entry, "Knocked down try-works." This innovation marked a tremendous advance in the fishery. Greenland whalers were able to keep blubber in casks through the cold northern weather, but, until the installation of try-works on shipboard, the sperm whaler had been limited to short cruises of several weeks. A longer stay impaired the quality and value of the oil. In a very short while all American whalers were equipped with try-works.

The British Greenland Fishery continued to pack their blubber in casks through Scoresby's time, 1820, and Markham in 1875 describes the same process. It is not to be wondered that Greenland ships were always given a wide berth. The men while cutting-in stood on the whale with spikes on their shoes to keep them from slipping, and with their axes and knives cut out big chunks, which were hoisted one at a time to deck.

The American Fishery in the meanwhile had perfected the spiral method of cutting-in, a very much simpler and quicker process. The heave of the windlass turned the whale over and over in the water, and the entire blubber was removed in one continuous strip. This method was introduced by the Americans into the British South Sea Fishery at the time of the Revolution. But the British-manned ships of the Greenland Fleet did not adopt it.

GEAR AND CRAFT

The first cutting-stages were short fore-and-aft planks hung overside, one forward and one aft of the gangway. A man stood on each, with his back to the ship, and a breast rope lashed him to the bulwarks. This stage never entirely went out of fashion, as it was frequently required in "taking care of" the head of a large whale.

The "outrigger cutting-stage" came into general use in the sixties. This consisted of a long fore-and-aft plank boomed some ten or twelve feet from the ship's side, with two thwartship planks. The fore-and-aft plank had a waist-high handrail on the inside. The men were lashed to this rail and worked leaning over it and facing the ship.

About the most important single article of equipment aboard a whaler was the cask, and one of the most important men on shipboard and one who always commanded a good lay was the cooper, who frequently combined the office of cooper with that of shipkeeper. Practically every commodity in an outward-bound whaleship's cargo — provisions, slops, sails, coal for the galley stove, water, and whalecraft — was headed-up in casks. As this material was consumed or used, the casks were filled with oil. Those of the largest size would hold as much as fourteen barrels. They were made in two lengths, forty-six and fifty-two inches. The largest was forty-eight inches across the head, and weighed when full of oil about two tons. The smallest one, frequently no more than six inches across the head, and holding a mere barrel, was called a "ryer," possibly a corruption of Rider. Staves were made of one and one-quarter-inch white oak split by hand "with the grain," not sawed out as in ordinary cooperage. Until the end of the industry, every whale-oil cask used in New Bedford was made in an old-fashioned coopershop entirely by hand. The earlier casks were wooden-hooped, but the first American whaler in the Pacific, the *Beaver* of Nantucket in 1791, carried four hundred casks, iron-hooped, and fourteen hundred wooden-hooped, so it is evident that the iron-hooped cask has long been used. It is also evident, from the number the *Beaver* carried, that large casks were not employed at that time. When big casks became the rule, a ship's complement of all sizes was between four and six hundred, averaging five to seven barrels each, according to the size of the ship and the convenience of her hold. Many of the casks went aboard ship "broken down." The staves were in "shooks," which were iron-hooped bales, and the heads and hoops were stowed in casks. The cooper's work on shipboard was constant; casks had to be opened and "flagged" — that is, rushes were placed in the seams to provide a sort of caulking which prevented leakage. Shooks were broken out and casks set up, and frequently new ones made.

It is said that at one time there were sixty cooperages in the New Bedford district. An astonishing number of articles of ship's furniture bore evidence of the cooper's trade. In the whaleboat were line-tubs, water-breaker, a piggin for bailing, and a lantern-keg, which latter contained lantern, candles, matches, flint and steel, tobacco, and hard bread, and was to be "opened up" in case the boats were out overnight. On deck were mess-kids, mess-buckets, deck-buckets, harness-cask, slush-tubs, wash-tubs, and coolers. Each watch had a five-gallon firkin for molasses. There were mincing-tubs, and a tub at the waist which was filled with a draw-bucket from overside when flushing decks. There was a case-bucket with a pointed or round bottom to be forced into the case in bailing. The cooper's grindstone had for a base an elliptical-topped tub, and the cooper's devil, or anvil, was set on a substantial truncated cone made of staves and hoops.

In the lower hold the casks were stowed on their sides, fore-and-aft, in two tiers. The lower was called the "ground" tier, and the upper-tier casks were termed "riders." They were stowed either "square tier," all square, or "bilge to cuntline"; that is, the second tier, or riders, were "staggered" to fit into the interstices of the ground-tier casks. They were secured, stowed, and wedged into place with "dunnage," the sailor's name for cordwood. Between-decks, where provisions and stores were kept, the casks were upended.

After oil was stowed down, it was necessary to pour water over the casks three or four times weekly, to keep them tight; otherwise they would shrink and oil would be lost. So for an hour or so every day the pumps wheezed; one day the water was pumped in, and the next day it was pumped out again.

To raise and lower huge casks through the hatches in a sea-way required considerable skill. It was done every day in breaking-out, but so far as was possible oil was piped through a leather hose to casks already stowed. In hoisting, casks were handled with can hooks which gripped the chines, and were far safer than slings. Special hooks were required for the huge casks of a whaler, the bearing edges of which were six or eight inches long, enough to span the chines of two staves.

With these hooks the stevedores broke out the cargo and emptied the hold at the end of the voyage, employing in the last few years of whaling a donkey-engine on a lighter between the ship and the wharf. The casks were hoisted and swung directly to the wharf on a derrick arm. This process of discharging was the final development in the methods of the Yankee whalefishery.

SUNBEAM LOWERING BOATS
In the collection of the New Bedford Public Library

STRIPPING THE WANDERER, 1923
Painting owned by Mr. Stanley Clark

WHALER AND BUMBOATS AT BRAVA
Painting owned by Mr. Nat C. Smith

THE GREYHOUND IN PORT, 1918
Painting owned by Mrs. E. D. Pouch

BEFORE THE MAST
A forward deck view of the Charles W. Morgan

THE GREYHOUND: WITH SAILS BENT PREPARATORY TO SAILING
Painting owned by Mr. Andrew G. Pierce, Jr.

GRAY FOG
Painting owned by Mr. Everett Fabyan

THE WANDERER WITH SAILS CLEWED
Breaking out oil

SPERM WHALING
Sketch for a lunette in the American Museum of Natural History

BOWHEAD WHALING
Sketch for a lunette in the American Museum of Natural History

WIND FROM THE NORTHWEST
Painting owned by Mr. George H. Taber

THE CAPTAIN: THE SUNBEAM'S AFTER CABIN
Painting owned by Mr. Henry S. Pyle

THE GREYHOUND AND OTHER CRAFT
Painting owned by Mr. E. D. Pouch

WRECK OF THE WANDERER AT CUTTYHUNK
August 26, 1924
Painting owned by Mr. R. Eugene Ashley

THE SUNBEAM CRUISING
Painting owned by Mrs. Joseph L. Woolston

SCHOONER A. E. WHYLAND AND MERCHANT BARKENTINE SAVOIE

HARBOR MIST
The Charles W. Morgan at Kelley's Wharf, Fairhaven

ANCIENT CRAFT, 1914

At the left are the sterns of the Nicholson (schooner) and the Morning Star (bark). The barkentine in the middle distance is a merchantman. In the foreground is the bark Charles W. Morgan with "ice-breaker" bows of more than four feet thickness of solid oak. (This reinforcement was afterward removed.)

THE BARK SWALLOW
Overmantel in the house of Mr. Arthur Delano

THE WANDERER IN WINTER QUARTERS
Painting owned by the Morse Twist Drill and Machine Company

THE CHARLES W. MORGAN AT FAIRHAVEN
Painting owned by Miss Emily M. Hussey

A WHALESHIP ON THE MARINE RAILWAY AT FAIRHAVEN
In the permanent collection of the Brooklyn Museum

A WHALER OUTWARD BOUND
Painting owned by Mr. Lammot du Pont

THE COOPER SHOP: FIRING THE CASK.
In the collection of The Mariners' Museum, Newport News

THROUGH THE STRAITS
Painting owned by Mrs. Frank S. Wilcox

FAIRHAVEN

THE SUNBEAM IN A FOG
Painting owned by Dr. Edwin P. Seaver, Jr.

CHAPTER X
THE WHALEMAN

> Our waist boat went down and
>   Of course got the start,
> Lay me on! Captain Bunker,
>   I'm Hell on the Dart.
>
> Now bend to your oars boys,
>   And make the boat fly,
> And mind just one thing now,
>   Keep clear of his eye.
>   *From "Captain Bunker," an old whaleman's song.*

THE people who brought whaling, one of the most sanguinary of all pursuits, to the highest degree of success were chiefly members of the Society of Friends; gentle people, opposed to violence in any form, who had been persecuted in the Massachusetts Bay Colony, and who had sought sanctuary at Nantucket. Among the islanders was also to be found a considerable strain of Pilgrim blood.

New Bedford was settled with exactly the same mixture of Quaker and Pilgrim, and when the whaling industry started in the younger town, many Nantucketers removed from the island to New Bedford.

These men and their sons made up the first whaling crews. As the business grew, the town grew, and as ships got bigger and crews larger, neighboring towns were called upon for their sons. Exactly as it had happened in earlier Nantucket, the enterprise in New Bedford became a neighborhood affair. One man made boats, another craft, a third supplied provisions, and often the payment for these supplies was contingent upon the success of the voyage.

Until whaling was well on the wane, it is safe to say that the majority of the crews were not only native Americans, but that they were also raised within easy distance of the port from which they sailed. James Templeman Brown said in 1887, "Captain Isaiah West, now eighty-six years of age, tells me that he remembers when he picked his entire crew within a radius of sixty miles of New Bedford, that oftentimes he was acquainted, either personally or through report, with the social standing or business qualifications of every man on his vessel; and also that he remembers the first foreigner, an Irishman, that shipped with him, the circumstance being commented upon, at that time, as a remarkable one."

In the early shore fishing of Nantucket, it was found that Indians made excellent whalemen. The Nantucket sloops carried at least two and frequently five Indians for each boat. Right up to the end of whaling it was not unusual to find one or two Indian boat-steerers on a New Bedford ship. The Portuguese of the Western Islands were good boatmen and whalemen, and it became the custom to ship some of them, touching at the islands outward bound. The Sandwich Islanders were also excellent whalemen, and when the fishery in the middle of the nineteenth century had reached such proportions that it was difficult to get a full Yankee crew, these islanders were eagerly sought by the Pacific fleet. The merchant seaman made a very undesirable whaleman. The reason was almost purely a psychological one. His whole training had made him look upon a small boat as a last resort, and a flimsy one, in time of extreme peril. If he had been two or three voyages in the merchant service, nothing on earth could rid the sailor of his timidity in a small boat — he was no good whatever, except as a shipkeeper, aboard a whaler. And when he was relegated to that job, as invariably he was, his pride was hurt, and he became a malcontent.

When the Civil War broke, whaling practically ceased. Massachusetts furnished 1226 naval officers out of a total of 5956 in the Union Navy, almost twenty-one per cent. (Starbuck, p. 113.) A goodly proportion of these officers were whalemen, and among the seamen of the Navy were to be found most of the Yankee whalers' foremast hands. When the war was over, it was evident that the day of the whaler had past. The Yankee whaleman turned to other pursuits and he never again dominated the forecastle.

In 1887, J. Templeman Brown states, "Few Americans below the rank of mates and captains are to be found on whaling vessels now sailing from our ports."

There were no classes of midshipmen in the whalefishery to teach young men to become officers. The school for whaling officers had been the forecastle and the steerage. When fiction writers speak of the officers of the whaling fleet as a race apart from the crew, it is evident they have missed one of the salient facts of the fishery.

I have known many fine old men who have passed from one end of a ship to the other, foremast hand to master, and if the caliber of the men turned out is an indication, it was a pretty good school. Captain William I. Shockley, whom I have quoted frequently in this volume, sailed before the mast at the age of fifteen. The voyage lasted forty-five months. He was made boat-steerer eleven months out, and he struck fifty-five of the one hundred and five whales his ship took on this his first voyage.

I know of another captain who sailed as cabin-boy at the age of fourteen. He grew so rapidly and became so awkward that he was "driven" to the forecastle. Four years after he sailed, having arrived at the mature age of eighteen, he returned to New Bedford weighing one hundred and seventy pounds, standing six feet in his socks, and a full-fledged boat-steerer to boot. He entered the door of his mother's kitchen just as the good woman was drawing a pan of bread from the oven. When his fog-horn bass boomed out, "Hello, Ma!" his mother dropped her pan, took one look, and promptly fainted away on the kitchen floor.

The whaling master then was a man who had learned his trade from the bottom up. Before the mast he had the sons of his neighbors, his relations, his wife's relations, and his fellow church-members. His was a double responsibility; one was to the ship, the other was to the man before the mast. He was no easy disciplinarian. It was his task to make men and whalemen out of the raw youngsters who generally made up the bulk of the foremast hands. Probably they were eager and willing at the start; but if a bit spoiled and allowed to get under the influence of "forecastle lawyers," they might easily present a problem. The fact is that the attitude of the captain of a whaler toward his crew was paternalistic to a marked degree. A merchant captain's responsibility ended when he safely took his ship from one port to another. But the whaling captain's real duties began when, once at sea, it became his business to find a cargo. In the pursuit of this cargo he visited desolate and dangerous seas, and among these for a matter of several years he was responsible for the health and happiness of his crew and the safety of his ship.

On a four-year voyage with a crew of forty hands, one hundred and sixty life-years were consumed by men engaged in exceedingly hazardous work. So it was inevitable that death would occur on an average voyage. The captain, without teaching, and with only experience and a medical book to guide him, had to diagnose symptoms and dispense medicine, set limbs and perform operations. He had the legal power, with its attendant responsibility, to punish crime aboard ship. He had to bargain for stores at the islands he visited, and to ship additional crew when needed. Frequently he had to assume entire charge of extensive repairs to his ship, heaving her down if necessary in some isolated port.

Seamanship was the least part of a whaleman's business. It was oil that sent him to sea, and whaling was a highly technical business, the successful pursuit of which required both natural aptitude and constant application. Nevertheless old whalemen were famous seamen. Nantucketers were noted for their navigation at a time when their only available instruments were quadrant and lead-line. American whalemen charted over four hundred islands in the South Seas. They first discovered the fact of the Northwest Passage, half of all Polar discovery is to their credit, and the first chart of the Gulf Stream was made by Captain Timothy Folger, of Nantucket, at the behest of Benjamin Franklin, a relative of his, who presented the chart to the merchants of London.

Whaling was a life that required certain high qualifications. It may be taken for granted that the men who were attracted to it were beyond the average in physical courage, or they would not have planned to face in frequent combat the greatest animal that ever lived; and that they were beyond the average in resolution and ambition, since they were willing to renounce, for years on end the food, the comforts, and the amusements of shore, and the companionship and society of friends, wives, and children; and to endure hardships and labor unremittingly, often against overwhelming misfortune, until the purpose of their voyages was accomplished.

In the duration of a long whaling voyage there came inevitably a time when no whales were taken or sighted for weeks, or even months, when apparently there was none left in all the ocean; when one day, no different from all the rest, succeeded another that had seemed interminable; a time when home and the end of the voyage were still so far away that it was futile to look forward to them. The same food, the same faces, the same sea, and the same routine, day after day without end, and not one barrel of oil nearer "full ship." The call of the watch, the ring of the ship's bell, the hail to the masthead, the "all's well" of the lookout, only served to punctuate the endless round of sameness. Nerves began to fray; friends began to look askance at one another, to interrupt familiar songs and yarns. Every man knew the uttermost thoughts of all his shipmates; there was nothing further of interest to be heard from any one of them. Silence reigned in the forecastle, and all hands brooded over misfortunes and wrongs that rapidly magnified themselves into formidable proportions. This was the time to bring out the leadership of the officers. The situation was one that it was their duty to anticipate. It was their business to protect the men from themselves, and to guard against any untoward circumstance that might fan the general moodiness into an active flicker of evil.

At such a time the appearance of anything on the horizon, a school of porpoise or a passing merchant ship, was a Godsend. The most trivial happening provided interest and conversation for days, so scant had become the grist of their minds. The books and magazines aboard were already dog-eared and committed to memory; the mere sight of them was hateful. On Sundays the men no longer shaved or cut their hair, long beards drooped from their chins, clothes became ragged and shed their buttons unnoticed.

The officers kept the watches occupied as never before; spars were scraped and slushed, rigging was overhauled, boats were patched and sails were mended. This was the time for the practice of scrimshaw, and the absorbing interest of the sailor in this occupation was often his salvation. Another fruitful field of interest was ornamental knot-work, a handicraft at which the whaleman excelled. He made elaborate beckets for his sea-chest and lanyards for his clothes-bags. At his officers' behest, he fashioned deck-fittings for the ship, bell ropes, fringes for line-tub covers, man-ropes and decorative draw-buckets.

The excessive length of whaling voyages has always been the outstanding fact of a whaleman's life, the fact that marked him apart from all other seamen. To the layman it was an appalling thing, and it kept away from the service all but the hardiest and the most venturesome. The longest whaling voyage on record, that of the *Nile* of New London, lasted nearly eleven years. She sailed in May, 1858, and returned in April, 1869. In 1865 the bark *General Scott* returned to Fairhaven after a four years' cruise with all her original officers aboard. This was accounted a very unusual circumstance.

A six months' North Atlantic voyage, officially called a "'tween seasons" voyage, unofficially called a "plum-pudding voyage," was much scorned by the sturdy old-timers of New Bedford. Such a voyage was generally carried on by Provincetown brigs and schooners. One old New Bedford skipper, who had been coerced against his inclination to make such a voyage, was busily engaged casting off from the wharf when his agent approached and whispered in his ear, "Captain Jones, you've forgotten to kiss your wife good-bye!" Without shifting his gaze from aloft, the captain demanded — "What's ailin' her? I'm only going to be gone six months."

There have been many yarns told illustrative of the extreme length and precarious nature of a whaleman's voyages. One captain reported upon his return from a four years' fruitless search that he hadn't a single barrel of oil or a single pound of bone aboard, but he'd "had a damn fine sail!" A California clipper is said to have once hailed a whaler in the neighborhood of the Horn and to have asked, "How long from port?" One of a row of ragged and long-bearded men, who were lined up at the rail, answered, "We don't remember, but we were young men when we started!"

There was one practice among whalers designed to relieve the tedium of the voyage for both officers and men which had no counterpart in other branches of service. This was the "gam," a sort of visit or truce between two or more ships of the fleet. Two whalers meeting at sea would set signals and heave to. In a short while the captain and a boat's crew or even half a watch would leave one ship, and from the other the mate and an equal number of men would set forth. The visit might last a day, a week, or even longer. The two ships sailed in company; and when whales were sighted, they were frequently hunted in common, in which case the ships were said to be "mating." The amount of the catch was equally credited between the two ships. Frequently one ship had letters for the other, or she bore word that such and such a ship carried letters, and that this other ship might be expected on certain whaling grounds at some specified season. As the gam continued, other groups of men were transferred, until every hand of the two ships had the benefits of a change of scene and a change of cookery, and met all the men on the other ship, and generally had an opportunity to readjust his viewpoint. It is my experience that, in spite of the relief from monotony, most sailors were glad when a gam was over and they could settle down again to routine. But it served a useful purpose; it made a break, and the men were left more contented and anxious to be about the business of their voyage.

Desertions from whalers were frequent. A whaler made port, if it were possible and expedient, every few months, to secure fresh provisions and to give the crew a run ashore. If a tropical island were reached a few months out, and a young sailor had a chance to stretch his legs on the beach, he immediately was tempted to desert. It was the most natural thing in the world. He had been actuated by a spirit of adventure when he signed for the voyage: for the first time in his life he was landed in an exotic place with palm trees and tropic light and easy women to distract him. He had only a dollar or so in his dungarees, but if he had been forehanded he had laid in a supply of tobacco and slops, and these were currency anywhere. He knew that the next ship to come along would be glad to ship him in place of those who would desert from her. He knew that the nearest consul was bound to feed him until he could be sent home at government expense. Perhaps because of his own stupidity or stubbornness, he was at odds with one of the officers; perhaps there was a bully in the watch who had made his life miserable; or he had turned out to be the butt of all the forecastle jests. The voyage was young, the amount of money due him was inconsiderable. What more natural than that he should make for the interior and hide until he saw his ship sail away? There was no legal penalty attached to desertion, and if he was caught and returned to his ship, his officers would not harbor it against him, for the chances were that they had each in the past done the same thing. An anecdote, told by the late W. W. Crapo, well illustrates the irresponsibility of the sailor. A well-known New Bedford skipper was impeached by Father Damien of Molokai. "Captain, there *must* be truth in these stories of ill treatment, else so many men would not desert!" "Father," said the captain, "if I cleared my ship for Heaven tomorrow, and touched at Hell next month, every damn one of 'em would desert if I gave him a chance."

In the *Sunbeam's* voyage, 1904–06, five men deserted at one island. The whaling had been good, the food was the best possible considering the nature of the voyage, and there had never been any ill-treatment. One of the older foremast hands came aft and told the mate that a certain boy was preparing to desert. The boy had flinched in the boat and his crew wasn't proud of him. "Might's well let him go," said Mr. Smith. "He's too young," argued the man, "and he comes from a good family." "Mebbe you're right," conceded Mr. Smith; "I'll speak to the captain." So the boy was discussed in the cabin, and the only point weighed was, "What is best for him?" It was decided that the boy should stay. So his desertion was postponed several months.

In fiction the typical whaling officer was a man-eating ogre, but in real life he was generally the affectionate father of a rather small family and was distinguished from his equals ashore mainly by his habit of long silences, which he could not shake, even when off duty. The reader should not accept without reserve the hectic pictures of whaling by Bullen, Browne, and others. And authors of books purporting to be serious maritime history have been as flagrantly at fault as the story-writers.

Young men were sometimes advanced too rapidly, were made officers when they were quite immature; and then under heavy weight of responsibility nerves may have snapped. To circumstances of this sort can be attributed several dark pages in whaling, such as the mutiny on the ship *Globe*, where several persons were murdered, and not a man on shipboard was more than twenty-two years old. The history of the late war shows what boys without perspective will sometimes do under stress. There were undoubtedly cases of brutality. In an industry the size of whaling, all kinds of men would be found. But the outstanding qualities required in a good whaleman were not the characteristic qualities of a coward and a bully.

Now and then a man who went whaling came whimpering home to write a peevish account, full of spleen and bewilderment, concerning a life he was not fitted to comprehend. His account is usually characterized by frequent error in the descriptions of the whale-hunt; although the descriptions of deck operations are reasonably accurate. He handled his pen better than he did his oar. If this same individual had been indentured on a New England farm, or had been put to work on a construction job, and had not been allowed to quit until his time was up, his autobiography would doubtless have carried the same spiteful whine and would have disclosed the same hatred for the men whose duty it was to see that he did not shirk and leave his work for better men to do. No doubt he painted whaling as he saw it, but his viewpoint was the viewpoint of the square peg.

Among the most widely read whaling books by non-professional whalemen are "Etchings of a Whaling Cruise," by J. Ross Browne, and "The Cruise of the Cachalot," by Frank T. Bullen. Both of these are stirring and graphic yarns, but they are not to be taken as accurate descriptions of American whaling. In the latter book the technical terms and expressions are largely borrowed from the British Fishery.

Such words as "Cachalot" for Sperm Whale, "flenzing" for cutting-in, "gallows" for skids, "wheft" for waif, "crow's-nest" for hoops, were not used on American ships. A crow's-nest in the American fishery is a shelter from the wind used only in the Arctic. Neither of these authors is at his best when describing the work of the boats while down for whales. There is one writer of whaling fiction whose book may be taken seriously and unquestioningly. There could be no truer picture of whaling or finer story of the sea than Herman Melville's "Moby Dick." Melville knew his subject, and if he occasionally borrowed an English term, his was obviously the virtuosity of the scholar and not the ignorance of the novice.

The food of the whaler has always been a ripe subject for vilification. The exigencies of the long voyage limited fare to provisions that could stand extremes of temperature and last for protracted periods. The ship agents of New Bedford bought the best material available. To do anything else would have been to waste money, for only the best stood any chance of lasting a full voyage. With good care, good luck, and a good cook there was no reason why a whaler could not have had wholesome fare. But there was always danger of food getting wet and mildewing; and good sea-cooks unfortunately were rare.

A booklet entitled "Memorandums, Tables and Schedule of Articles Comprising the Outfit for a whaling Voyage," which was required as a checklist in fitting out a whaler, contained the following list of eatables considered essential for a voyage. There was also an open page for such additional entries of canned goods and stores as were fancied by the agent or captain:

Flour, baked	Lemon syrup	Rice	Cabbages	Allspice
Flour, packed	Pickles	Dried apples	Souchong tea	Cloves
Kiln-dried meal	Mackerel	Raisins	Chocolate	Cinnamon
Mess beef	Tongues and sounds	Beans	Crushed sugar	Saleratus
Prime pork	Codfish	Peas	Mustard	Pepper sauce
Pork hams	Sugar	Corn	Black pepper	Table salt
Molasses	Butter	Potatoes	Cayenne pepper	Sweet oil
Vinegar	Cheese	Onions	Ginger	Coffee

Lime juice, wines, brandy, gin, and rum were included under "medicinal stores." The food of the cabin was identical with the food of the forecastle. When there were vegetables aboard ship, the three messes, cabin, steerage, and forecastle, shared them equally. The crew habitually grumbled over the same food that the officers ate without complaint. This is characteristic of seamen the world over, and it meant nothing, save that there was a dearth of subject-matter for conversation in the forecastle. Since the officers were older men who had served their apprenticeship, it would seem that they were entitled to a better table, but the only difference that I ever saw between the cabin and forecastle fare was that in the cabin food was served on a table and included butter and a few condiments. Not infrequently the forecastle also had butter. But butter is an ephemeral dish, and its value aboard ship is easily overrated. To quote from Davis: "When butter has twice crossed the Line, and has reached the age of two years, no stomach can accept it as butter, but as cheese it seems quite mild."

In the forecastle mess each man would sit on his own sea-chest in front of his berth, and with the aid of his sheath-knife and fingers would eat from a small, deep tin pan which he held in his lap. All his food was dumped into the one dish. He also had a pint tin measure, called in nautical parlance a "pot," which was filled with black, steaming coffee. Each watch was allotted so much molasses a week or month, and this served to sweeten coffee. The men had all the hard bread they could eat at all times, and the hard bread of the whalers has always been the best carried in any service.

Jenkins, speaking of food in 1801 (p. 202), states, "The feeding conditions in merchant vessels generally were at this time extremely bad, and it does not appear that whalers [whalemen] were much worse off than other sailors — probably the whalers [whalemen] were, if anything, rather better off than the average seaman." But in the last days of the fishery, undoubtedly the food of the merchant and naval services was preferable. Methods of refrigeration had been perfected, and other ships making ports at shorter intervals were able to carry fresh food.

The true test of food was the condition of the men. In the last days of the whale-fishery, although two-year voyages were the rule, the one great curse of the protracted voyage, scurvy, had been eliminated. Crews generally were healthy and strong, and except for an occasional case of beri-beri, there was little chance of disease being contracted aboard ship.

My own experience was this: I found the food of the *Sunbeam* unattractive, but ate it without discomfort and when we made port discovered I had gained ten pounds in weight. I have known old whalemen to have such a longing for sea-fare that they have gone to the dock to meet an incoming ship in order to get an ancient piece of salt-horse to take home to be cooked.

The whaleship "agent" of New Bedford combined the offices of shipping-master, outfitter, and owner. In this way he was obligated to himself to fill each office fairly, since the skimping of one job resulted in injury to himself in one of his other capacities. He shipped the best men and officers he could obtain, and bought the most suitable provisions, since this was the best insurance for an economical and successful voyage.

A ship was divided into a number of "shares," also called "pieces" or "parts." The agent generally owned the controlling interest, although he was sometimes merely retained to oversee the outfitting and attend to the settling of the voyage. Because of the extremely hazardous nature of the calling, the old Yankee agent seldom owned a ship outright. A wealthy agent might own a single ship and a controlling interest in four or five more, and then decreasing amounts in a dozen others. In this way he covered his losses unless a season proved very disastrous indeed. A ship might be divided into any number of shares thought fit, eight, sixteen, twenty-five, thirty-two, or maybe sixty-four. Usually the captain was asked to "buy in," thereby adding to his responsibilities the certainty of loss if the voyage were unsuccessful. The owners shared the losses *pro rata*, and the fact that he was part owner made the captain cautious in expenditure, and he could be depended on to recruit as economically as possible, since he stood to gain or lose at both ends.

Many New Bedford agents maintained their own stores and outfitting establishments. Outfitters were always referred to locally as "sharks." In spite of this name, they conducted the cheapest shops at which to buy their own sort of goods. Their range of stock was not wide, and the turnover was tremendous. They knew what a sailor really needed, and although this was often at variance with a "greenie's" preconceived idea, he had to take what was handed him. Without doubt the sailor sometimes paid dearly for his outfit, when the outfitter was at the same time the agent, and thrift got the better of him. As for the price of clothing aboard ship, slops were carried for three and four years, and the stock was "turned over" just once. A shop ashore plans to "turn over" its stock six or eight times in that period. There was bound to be a certain loss from deterioration, resulting from dampness, rust, rats, etc. There was also a loss if a sailor ran away with his outfit near the beginning of a voyage before it was paid for. Any profits from the slop-chest went into the gross of the voyage, and the sailor got his proportionate share when the voyage was settled. On the whole, the prices appear to have been fair.

Another task of the agent was to secure crews. In the early days of whaling the whalemen were "raised on the spot"; all-American crews gathered from the immediately surrounding country. But beginning with the decline of the fishery and its attendant loss of glamour, the youth of the countryside turned elsewhere and the character of crews changed. For some time it was the custom to recruit at the Azores and the Sandwich Islands, and these islands furnished large numbers of men.

The Cape Verdes were the most fruitful source in later years. The Brava, as the Cape Verde Islander was called, was black as an African, but had straight or nearly straight hair. Presumably he was a mixture of Moorish and African blood with a dash of Portuguese. He was more energetic than the African and proved an excellent whaleman. As it became more and more difficult to secure crews, the agents became less particular. After engaging enough whalemen to man the boats, they filled up with anything that was available. They figured that a whaling cruise would either make or break a man. In the latter case he would desert at the first opportunity, and being of little account, his place was easily filled.

A whaleman received no wages. His reward for labor directly reflected the quality of his work, since his "lay" was a proportionate share in the proceeds of the voyage. Sometimes whalemen received very little indeed. But a successful voyage made a fair return when we consider that the men were housed and fed for the duration of the voyage and that agents paid all bills. A successful voyage might mean over $100,000 to be divided between owners and crew. Four-year voyages of around $300,000 have been settled when sperm-oil sold at a dollar a gallon and bone had a ready market.

The following list will indicate the average lays allowed aboard New Bedford ships in recent years:

	3-Boat Ship	4-Boat Ship
Cabin-boy	215 (i.e. 1/215th)	225
Green hand	170–180	160–190
Able seaman	125–150	140–160
Boat-steerer	65– 80	74– 90
Fourth mate	50– 60	60– 65
Third mate	38– 45	45– 60
Second mate	28– 35	30– 40
Mate	18– 23	20– 25
Cooper	45– 55	50– 60
Cook	125–150	140–160
Steward	60– 80	100–150
Blacksmith	140	150
Master	8– 16	8– 18

If a man for any reason did not complete a voyage, his lay was figured upon the relation the period of time he had served bore to the whole length of the voyage, and had nothing to do with the number of whales actually taken while he was aboard.

Success in a whaling voyage was perhaps more dependent than any other commercial undertaking upon the spirit that was brought to the job. Scoresby points out that, whereas at the beginning of the Spitsbergen Fishery, "the Dutch far surpassed all other whalemen, by the end of the eighteenth century they were very indifferent ones." He rightly concluded that this was due to their loss of keenness. To sight whales, in the first place, required an eager alertness that only the born sportsman possessed, and to capture the whale afterwards required an untiring energy and an absolute preparedness. Any half-heartedness spelled defeat for the whole enterprise. Whaleboat crews as a general thing worked like well-kept machines.

Sometimes through bad luck a crew lost its morale. The men at the masthead became lackadaisical and not one half the number of whales were raised that a conquering vessel would have found in the same waters. This condition would be recognized by the owners; new officers would be sent out, some of the men would be transferred to other vessels, and the whaling grounds changed, to give the ship a fresh start.

The telltale signal of the whale is the faintest puff when shown against a far horizon, and it lasts but a moment. Unless the observer were expecting that signal, believing in his luck, and knowing that the sighting of a whale would mean actual money in his own pocket, he was not going to find it. When whaleboats were lowered away, the ensuing battle was not a simple matter of courage and skill. Success depended largely upon sheer ferocity of attack, of total submergence of all thought except that of getting into an advantageous position at the earliest possible moment, and of getting fast ahead of any other boat. It was the spirit of competition between boats that was at once the main secret of successful whaling and the cause of the greatest annoyance; "whaling for glory" it was called. For aboard a successful ship the competition at times would grow so keen that boat-crews overlooked the fact that every whale captured benefited all alike. They raced and jockeyed for position, and in a close finish, with boats jammed together at the flank of a whale, have been known deliberately to foul one another; to dart harpoons across each other's boats, imperilling both the boats and the lives of all concerned, and then to ride blithely off, fast to the whale, waving their hands or thumbing noses to their unfortunate comrades struggling in the water. The sport was the thing: they were a high-tempered lot, the old Yankee whalemen, and their motto was, "A dead whale or a stove boat!"

OLD NEW BEDFORD

Merrill's Wharf Loft Building (burned 1925) and the Bark Greyhound
Painting owned by Mrs. F. Gilbert Hinsdale

SAILING OF THE DESDEMONA
Overmantel in the house of Mr. Charles S. Ashley, Jr.

OCTAGON BUILDING AT FAIRHAVEN

In the center of this building stood a capstan around which was driven the horse that provided power for hauling whaleships up the railway
Painting owned by the estate of Mr. Warren Delano

COOPERING CASKS FOR THE GREYHOUND
In the collection of The Mariners' Museum, Newport News

THE SWALLOW INSHORE OFF CUTTYHUNK
Owned by the New Bedford High School

NEW BEDFORD HARBOR
Three-masted whaling schooner Mystic drying sails
Painting owned by the Morse Twist Drill and Machine Company

BLUFF BOWS: THE CHARLES W. MORGAN

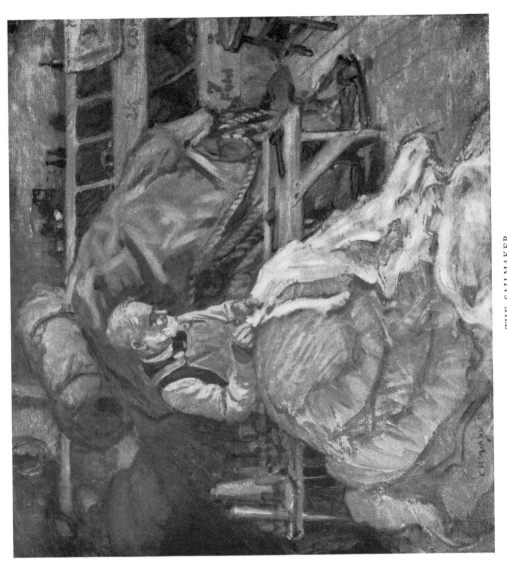

THE SAILMAKER
Painting owned by Mr. J. Franklin Briggs

THE CHARLES W. MORGAN FROM THE STATE PIER
Painting owned by Mr. George H. Taber

THE CHARLES W. MORGAN AT SEA
Painting owned by Mr. George P. Black

THE CHARLES W. MORGAN AT ROUND HILLS

This last New Bedford Whaler was saved through the efforts of Mr. Harry Neyland. Thirty-three people contributed to her purchase, and, after the aldermen of New Bedford had voted not to accept her as a gift to the city, Col. E. H. R. Green provided temporarily for her maintenance at Round Hills.

"A DEAD WHALE OR A STOVE BOAT!"
The hermaphrodite brig Daisy hove to with boats down
Painting owned by Dr. Robert Cushman Murphy

NEW BEDFORD WATERFRONT
Showing the merchant bark Charles C. Rice and the hermaphrodite brig Harry Smith
Painting owned by Mr. Joseph L. Woolston

THE SUNBEAM WEARING SHIP IN A FOG
Painting owned by Mrs. E. D. Pouch

THE OLD AND THE NEW

GETTING FAST
Painting owned by Mr. E. E. du Pont

THE GREYHOUND AT MERRILL'S WHARF

BARK WANDERER CLOSE-HAULED

THE SUSIE PRESCOTT AND THE CHARLES W. MORGAN
Painting owned by Dr. Curtis C. Tripp

THE SUNBEAM HAULED OUT, 1904

THE SHIP CARPENTER: AT THE SUNBEAM'S CUTWATER
Charcoal sketch

CAULKING THE SUNBEAM
Charcoal sketch in the collection of the New Bedford Public Library

MOUNTING AN IRON
Charcoal sketch owned by Mr. William Wing

"THERE SHE BLOWS!"
Charcoal sketch in the collection of the New Bedford Public Library

BOATS AWAY
Charcoal sketch in the collection of the New Bedford Public Library

SUNBEAM REACHING
Charcoal sketch

"A NANTUCKET SLEIGH-RIDE"
Charcoal sketch in the collection of the New Bedford Public Library

HOISTING BOATS: THE SUNBEAM, 1904
Charcoal sketch owned by Mr. Charles M. Hussey

THE FIRST BLANKET PIECE
Charcoal sketch owned by Mr. Eben Brown

MINCING
Charcoal sketch in the collection of the New Bedford Public Library

BOILING: THE SUNBEAM'S TRY-WORKS
Charcoal sketch in the collection of the New Bedford Public Library

CHAPTER XI

SCRIMSHAW

THE whaleman fortunately has left behind him one enduring monument. In his spare time he developed the only important indigenous folk art, except that of the Indians, we have ever had in America; the Art of Scrimshaw. From the very beginning of sperm-whaling, he began to fashion things of beauty and utility with his hands. St. John in 1782 says (p. 196), "I must confess that I have never seen more ingenuity in the use of the knife . . . in the many hours of Leisure which their long cruises afford them, they cut and carve a variety of boxes and pretty things . . . they showed me a variety of little bowls and other implements, executed cooper-wise."

Scrimshaw was a spontaneous growth made possible by the length and spare time of the long voyages, and its practice was so widespread among the ships that it may be said to have become universal. In fact, if we consider into what corners of the world the pursuit of their calling took the Yankee whalemen, it may not be exaggeration to say that there never has been another art so universal. In their spare time for a matter of seven or more decades, until Yankee sailors gave way to foreigners in the service, the better part of twenty thousand whalemen, year in and year out, spent the most of their leisure hours in seriously endeavoring to fashion something beautiful from the scraps of whale ivory, bone, tropic wood, tortoiseshell, silver, and any other fragments of material that the narrow confines of their ships could supply. Their isolation was such that it is not to be wondered that their design developed along original lines, and that the art which resulted proved unique. It is recorded that on some ships every man from captain down to cabin-boy had some article of Scrimshaw under way, and it once was a fiercely debated point among New Bedford owners whether the engrossing interest of the whaleman in his Scrimshaw was not seriously detrimental to the success of voyages. It was even alleged that on occasion men had sighted whales, and rather than be interrupted at some particularly fascinating point in the practice of their Art, had failed to announce them! Certain captains forbade Scrimshaw altogether; on some ships its practice was limited to the forecastle, and the man who brought his work on deck was liable to have it confiscated.

The origin of the word is unknown. Several of the dictionaries attempt to derive it from the surname Scrimshaw. But this is without doubt an error, since Scrimshaw is only the most recent of the several forms of the name. James Templeman Brown ("Fishery Industries of the United States," Washington, 1887) claims to have traced the word to Nantucket, and believes it to be of Indian origin. But this is merely surmise.

The earliest reference to the art, by name, appears to be in the manuscript logbook of the brig *By Chance* of Dartmouth, preserved in the collection of the New Bedford Whaling Museum. Under the date May 20, 1826, it reads, "All these 24 hours small breezes and thick foggy weather, made no sale [*sic*]. So ends this day, all hands employed Scrimshonting."

Cheever ("The Whale and His Captors," 1850) is the first author to give the word in print. He uses the form "skimshander." Melville, a year later, in "Moby Dick" speaks of "skrimshander." There is an analogy between these forms and the two synonymous words "skimp" and "scrimp" which suggests a possible derivation.

If we consider "scrimp" either as an adjective meaning "scant," or as a colloquial Yankee verb meaning "to economize," we have a word that may well be responsible for the first syllable of "Scrimshaw."

The nature of the material commonly used in Scrimshaw—the tooth of the Sperm Whale—makes scrimping necessary. The second mate doled out to the men, according to his whim or their requirements, ivory and bone. There was always a dearth of good material, for large pieces of whale ivory are rare. Whale's teeth are hollow at the root, and there never yet has been found one large enough to turn out a full-sized billiard ball. Waste was not to be thought of, for the officer kept record of what was given out, and hoarding was discouraged.

I have seen men swap tobacco, the universal currency aboard a whaler, wash clothes and do other menial tasks, in order to gain coveted pieces. I have watched them puzzle and plot, pencil-marking a veritable maze of lines which were to be followed in cutting up a tooth. The design of the scrimshawed article usually had to be suited to fit the form of the material at hand. A hacksaw would be borrowed of the cooper, and half a dozen men would stand around the work-bench, giving advice while a tooth was sawed.

Several writers on the subject have imagined that scrimshaw had its inspiration in the walrus tusk carvings of the Eskimos; but the work of the eastern Eskimo is very crude and the Arctic had never been penetrated by a whaler until 1835, so that the really excellent handiwork of the northwestern Eskimos was unknown to the whaleman at a time when the Art of Scrimshaw had reached full flower. Joseph Hart ("Miriam Coffin," 1834) describes the cane of Jethro Coffin: "It was wrought into diamonds and ridges, and squares and oblongs, like the war-clubs of the South Sea Islanders, and surmounted by the head of a grinning Sea Lion, with a straight black pin of polished whalebone driven through his ears." This might appear to suggest a South Sea Island origin. But the fact is, there is no need to hunt for an origin; the Yankee whaleman contrived objects for his own purposes, often highly original mechanical contrivances, embodied from his own design and covered with his own decoration. The source of his inspiration was to be found either in the home surroundings he had left or in his life at sea. There is little that is reminiscent of other arts.

The tools of Scrimshaw were generally knife, files, and saw. Melville says ("Moby Dick," New York, 1851), "Some of them have little boxes of dentistical-looking implements, especially intended for the skrimshandering business, but in general they toil with their jack-knives alone." The thin edges of the jawpan bone could be planed, for fresh ivory and bone are relatively soft. Many ships had home-made turning-lathes. But much fine turning was simulated with a file, and much of the pattern in Scrimshaw that resembles scroll-sawing was also file-work. Holes were drilled with gimlets made of nails. The countersinking for shell, silver, and mother-of-pearl inlay was scraped out with a knife, assisted maybe with a chisel. A grindstone under some circumstances was useful in smoothing both bone and ivory. The finishing was done with wood ashes, and a polish was added by rubbing with the palm of the hand.

If one man showed a better talent than the rest, his designs might influence the output of a whole ship. But the sailor is so independent a person that he usually preferred to be responsible for every feature of his own finished work. Within the single piece which his diligence might evolve in a voyage, he frequently evinced no mean skill in half a dozen handicrafts — joinery, turning, carving, inlay, coopering, engraving. The most familiar fruit of his craft were his graphics on Sperm Whales' teeth. What more natural than to wish to present a distant friend with a trophy of the whale-hunt, a huge tooth that, in actual conflict with a whale, had threatened him and now stood a symbol of his success? This tooth he smoothed with a file (the natural tooth presents a rough surface), and thereon drew carefully with his knife-point pictures of his ship and hair raising encounters between gigantic whales and himself and companions in their small boats. To make these lines visible, they were filled with black India-ink, with an occasional touch of red. When this had dried, the surface was polished.

There are no choicer bits of Americana lying around to be collected than these records of heroic deeds. If not in heroic mood, the whaleman pictured on his tooth the lady of his dreams. And it speaks well for his gallantry that in these pictures of women is the only instance where he felt his art inadequate, and borrowed freely from the illustrations in the magazines of the day. In all his other work is a marked and sturdy originality; here alone he frankly copied. But, after all, a tooth was a gift for a friend; the piece of Scrimshaw upon which he lavished his most sentimental regard, and limned his most inspired pictures, was the busk, since this was designed to be worn by his loved one next her very heart. It is perhaps necessary in these loose days to state that the busk was a flat fence-paling-like "stay" about two inches wide, which in the eighteenth and early nineteenth centuries was thrust into an open slit at the front of a corset. Any woman so fortified was bound to remain true to her sailor. Several of these are reproduced among the illustrations. Frequently they bore appropriate and tender verse. Two of these verses, which are from a busk in Mr. W. W. Bennett's collection, are typical, and well worth quoting, if only to show the skill of the whaleman in preserving a neat impartiality between the charms of his fair one and the lure of his calling:

> Accept, dear Girl this busk from me;
>   Carved by my humble hand.
> I took it from a Sparm Whale's Jaw,
>   One thousand miles from land!
> In many a gale,
> Has been the Whale,
>   In which this bone did rest,
> His time is past,
> His bone at last
>   Must now support thy brest [*sic*].

His design was often arbitrary, occasionally borrowed, but at its best it found its source in the life about him. His canes were spirally scored in imitation of rope. His box-covers were inlaid to resemble the radii of a compass card. The knots in the rigging; the stars in the heavens above him; the figurehead and stern-board of his ship; the fish of the sea; whales, birds, men, sails, boats, casks, bells; the wheel, the anchor, and other symbols of the sort, constituted his dictionary of ornament.

Although the graphic tooth was the commonest bit of Scrimshaw, and the busk the most sentimental, yet they provided no mechanical problems to tax Yankee ingenuity, and so may be regarded as the least of his efforts, in the gift of which he paid off his minor obligations. The fine fruit of Scrimshaw, the *magnum opus* of every scrimshanderer worth the name, the *chef d'œuvre* on which he was willing that his reputation should stand or fall, was his jagging-wheel. This was a gift for his wife or intended, no less. A jagging-wheel, it should be understood, is an elaborate implement for cutting, piercing, and crimping pies. What more natural than that the exiled Yankee should be haunted with thoughts of this culinary triumph of the New England kitchen, and far away at sea pay tribute to memory by fashioning for its creation a tool worthy of the task?

The historians of whaling have given scant space to the subject of Scrimshaw. They have been concerned with the business of whaling, and Scrimshaw was for the leisure hour. Scammon devotes a paragraph to it; J. Templeman Brown, two. Starbuck and Macy, both good Nantucketers, should have done better, for they knew the part that Scrimshaw had played in the daily life of the Nantucketer: but neither of them so much as mentions the subject.

The earliest dated piece of Scrimshaw that I am able to describe is a tooth in my own collection. It was decorated "off the coast of Japan" on the first voyage of the ship *Susan* of Nantucket, in 1829. Earlier dated pieces will possibly turn up, but there is nothing so early in the New Bedford Whaling Museum at present, nor in the several private collections with which I am familiar. The reverse of this tooth has an interesting couplet, which runs as follows:

> Death to the Living, Long live the Killers,
> Success to Sailors' wives, and Greasy Luck to Whalers.

And there is also a lively whaling scene, and the name of the master, Captain Frederick Swain.

Very soon after Nantucket's first Sperm whaling venture in 1712, his voyages having lengthened so that there was considerable idle time at his disposal, the whaleman busied himself with his jack-knife. Although at first wood was probably used more than ivory, it could have taken the whaleman little time to discover the possibilities of the choicer material.

The articles which he proceeded to make from bone and ivory are legion. There were bird-cages and baskets, work-boxes and ditty-boxes, checkerboards and dominoes, chessmen and jackstraws, swifts and reels, busks and stays, bodkins and knitting-needles, tool-handles and rolling-pins, clothes-pins and dish-mops, rings and bracelets, salt-shakers and napkin-rings, canes and whips, jig-blocks and belaying pins, coat-racks and embroidery-frames, writing-desks and boxes, cribbage-boards and work-tables, brackets and frames, rulers, penholders, paper-knives, brushes, butter-spreaders, cuff-links, scarf ornaments, fids, scribers, seam-rubbers, spool-racks, needle-cases, card-trays, sleds, baby-wagons, foot-scrapers, door-stops, hooks, knobs, and hinges.

Few ship models were built. This may have been because his quarters were cramped, or it may be that the whaleman saw a lighter beauty in the faster merchantmen that passed him by, and so failed to do honor to his own ship. Whatever it was, it is to be deplored. Whales he made a-plenty, and whaleboats quite a few. Any small household fittings of ornamental nature were made for which the absent swain and husbandman could find a purpose. He went to work at the sort of task he liked best, one that taxed his ingenuity and required a crafty hand. With loving care he wrought, and for the only time in his life gave free vent to whatever feeling for beauty was in him. To-day when we look at these things he created, we wonder to find that so much feeling actually was there, and that so much could be so deeply hidden.

If part of his work is clumsy and impractical, and misses its intent, nevertheless the beauty of the material imparts to it some degree of charm, so that the least successful of it is not to be passed over lightly, and the best of it ranks among the Fine Arts.

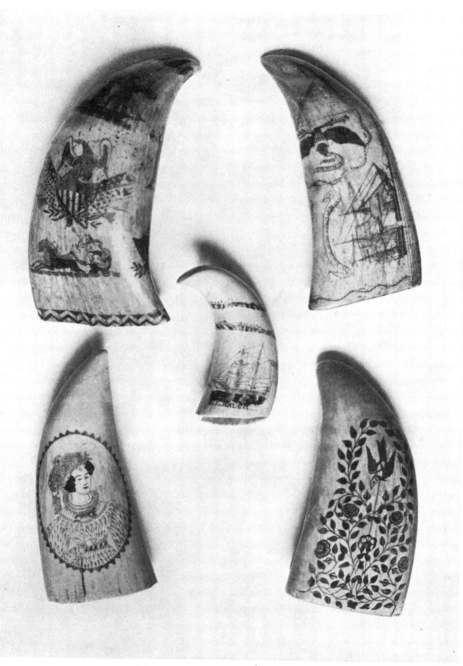
SCRIMSHAW WHALE'S TEETH
In the author's collection

SCRIMSHAW: SPERM WHALE'S TEETH
In the author's collection

SCRIMSHAW: TWO SPERM WHALE'S TEETH
The lower shows a whale being towed to the ship
In the collection of Mrs. F. Gilbert Hinsdale

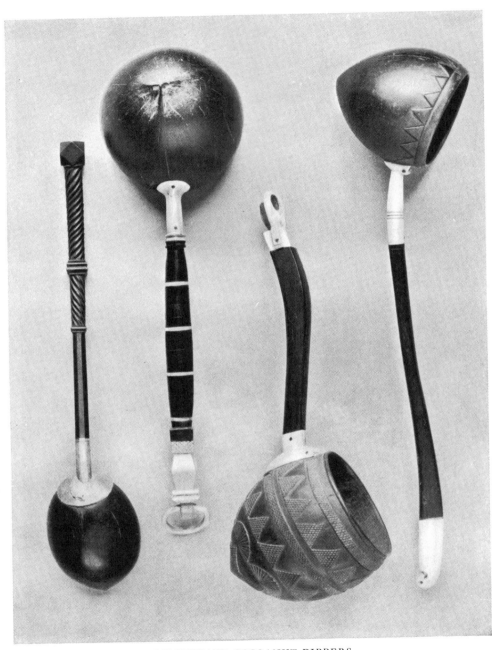

SCRIMSHAW: COCOANUT DIPPERS
From the collection of Mr. Edward F. Sanderson in the Nantucket Whaling Museum

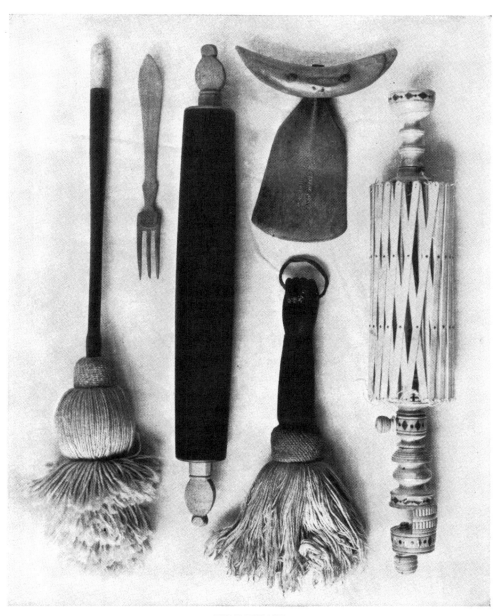

SCRIMSHAW
Mops, fork, rolling-pin, chopping-knife (made from a blubber-spade), and swift (for winding yarn)
In the author's collection

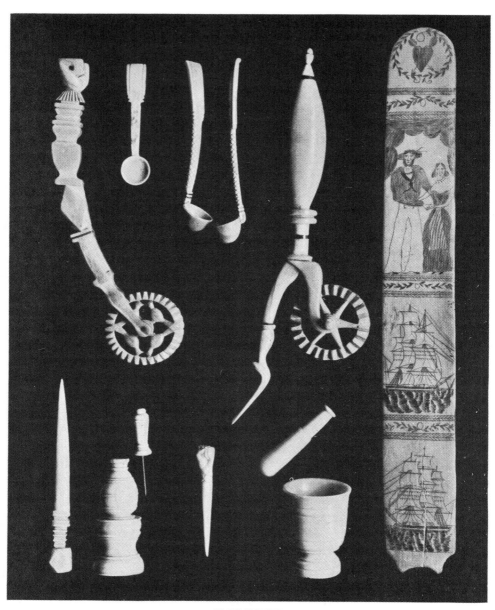

SCRIMSHAW
Jagging-wheel, salt-spoon, two mustard-spoons, jagging-wheel, busk, bodkin, pickwick, bodkin, and miniature mortar and pestle
In the author's collection

SCRIMSHAW BUSKS
The first is obviously the work of a homesick farmer boy on his first voyage; the others are more sophisticated in treatment and subject
In the collection of Mrs. F. Gilbert Hinsdale

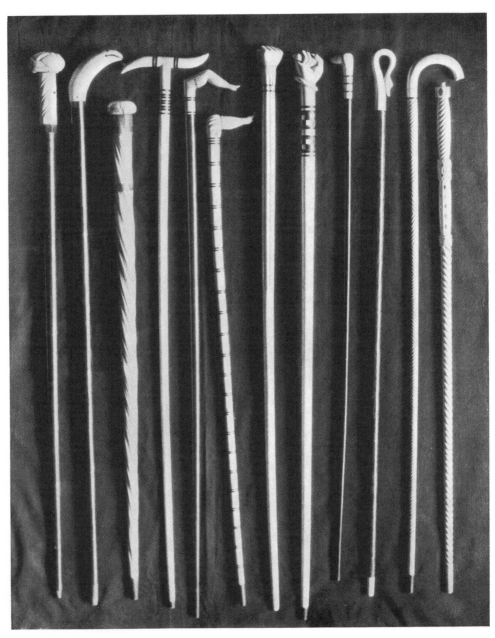

SCRIMSHAW CANES
The collection of Mr. Edward F. Sanderson in the Nantucket Whaling Museum

SCRIMSHAW: DITTY-BOX
In the collection of Mrs. F. Gilbert Hinsdale

SCRIMSHAW: A WHALING SCENE ETCHED ON PAN BONE
In the collection of Mrs. F. Gilbert Hinsdale

THE SECOND MATE'S DITTY-BAG

Most of these articles are scrimshaw. They include serving-mallets and serving-boards, awl, scriber, seam-rubbers, log-book stamp, knuckles, black-jack and cat-o'-nine-tails, fids, pricker, knife and sheath, needle-case, thimbles, sewing-palm, sailmaker's hook, pipes, and ditty-bag

In the author's collection

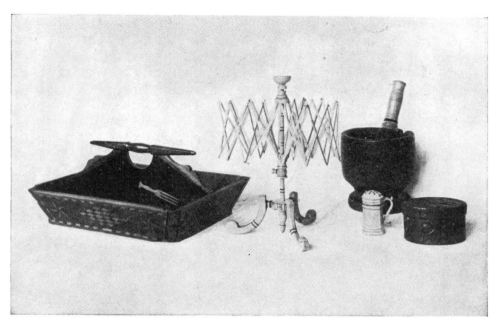

SCRIMSHAW: KNIFE-BOX, SILK-SWIFT, MORTAR AND PESTLE, SALT-SHAKER
AND DITTY-BOX
In the author's collection

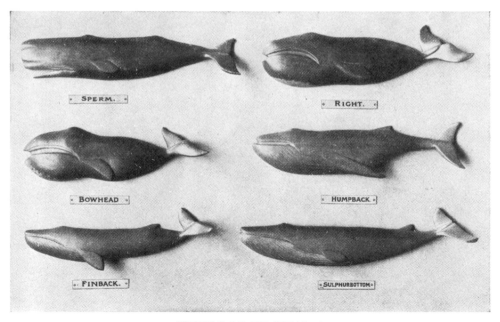

MODELS OF SIX WHALES
Made by Mr. Frank Wood

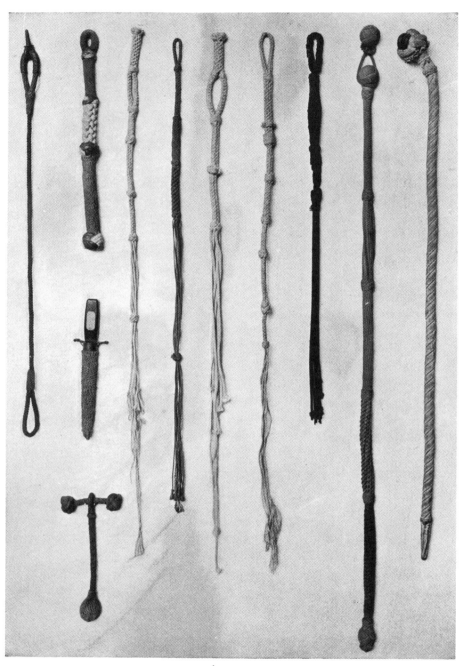

SAILORS' KNOT WORK
Boat gripe, bell rope, knife sheath, black-jack, five-bag lanyards, yoke-rope, and man-rope
In the collection of the New Bedford Whaling Museum

SIX SAILORS' CHESTS SHOWING ORNAMENTAL KNOTWORK AND NEEDLEWORK BECKETS
"The best type of sea chest had a sloping front, designed to save the sailor's shins when his ship heeled in a seaway. The best type of becket cleared the edge of the lid when lifted, thereby sparing the sailor's knuckles. Generally chests were lashed by their beckets, or else battens were nailed to the deck to prevent their shifting." Center left owned by Mr. Frank Wood, center right owned by Mrs. F. Gilbert Hinsdale, the remainder in the author's collection

FIGUREHEAD OF THE MARTHA, OF NEW BEDFORD
In the author's collection

BILLETHEAD FROM THE BARK ROUSSEAU
This ship was built for Stephen Gerard, of Philadelphia, in 1801 and was used by him as merchantman and yacht until sold to New Bedford and converted into a whaler. The ship was scrapped in the nineties.
In the collection of the New Bedford Whaling Museum

FIGUREHEAD OF THE MERMAID
In the author's collection

FIGUREHEAD OF THE BARTHOLOMEW GOSNOLD
In the collection of the New Bedford Whaling Museum

FOUR WHALERS' STERNBOARDS
Ship Jireh Perry of New Bedford (Collection of Mr. F. Gilbert Hinsdale)
Unidentified Wareham Ship (Collection of Mr. John Hall Jones)
Ship Reindeer of New Bedford (Collection of New Bedford Whaling Museum)
Bark Leonidas of New Bedford (Collection of New Bedford Whaling Museum)

CARVED AND PAINTED STERNBOARD OF THE SHIP MARY AND SUSAN, OF STONINGTON
Owned by Mr. Philip Sawyer

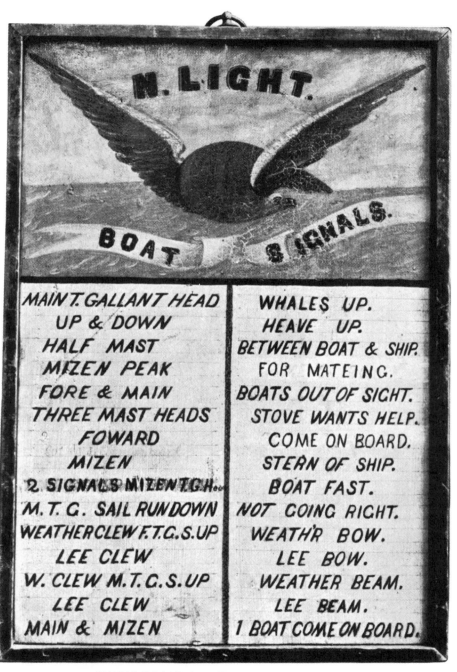

SIGNAL BOARD OF THE NORTHERN LIGHT
In the collection of Mr. F. Gilbert Hinsdale

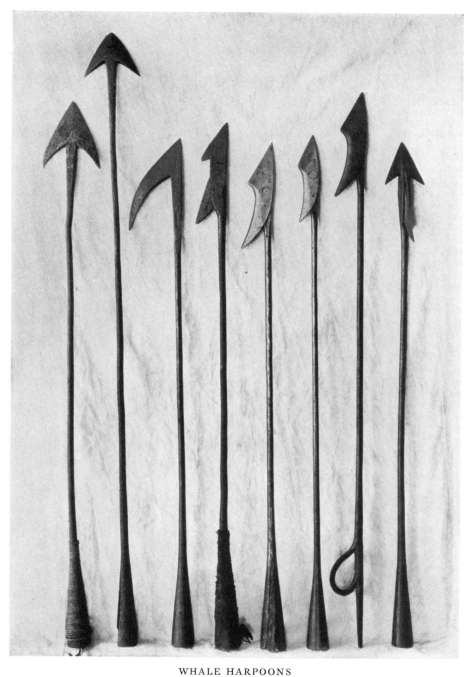

WHALE HARPOONS
Two-flued iron, long-shanked Arctic two-flued iron, single-flued iron, Temple's toggle, modern toggle,
Sag Harbor iron, darting-gun iron, toggle invented by Elisha Babcock (1860)
In the author's collection

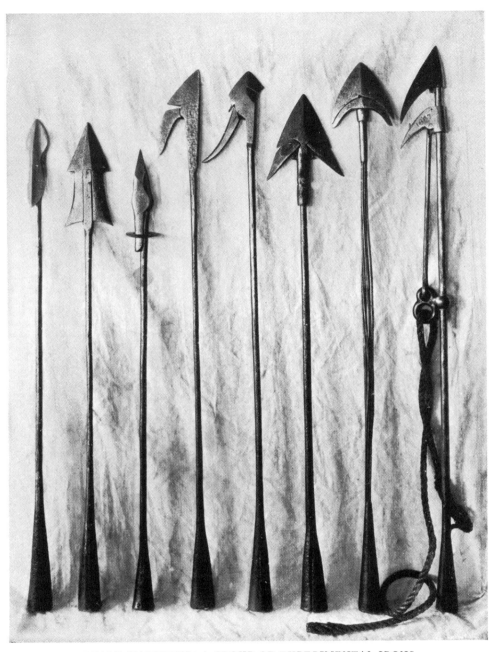

WHALE HARPOONS: A GROUP OF EXPERIMENTAL IRONS

Left to right: A balance toggle by James Durfee, N.B.; harpoon patented by George Doyle, Provincetown, 1858; grommet iron; hinged barb by Snow and Purrington, N.B.; hinged barb by J. M. Snow, N.B. (used successfully); removable head with two hinged barbs, patented, 1846, by Charles Randall, of Palmyra, Ga.; harpoon patented, 1846, by Holmes and West, of Tisbury, Mass.; the "heart-seeker," patented, 1857, by James Q. Kelley, Sag Harbor, N.Y.

In the collection of Mr. F. Gilbert Hinsdale

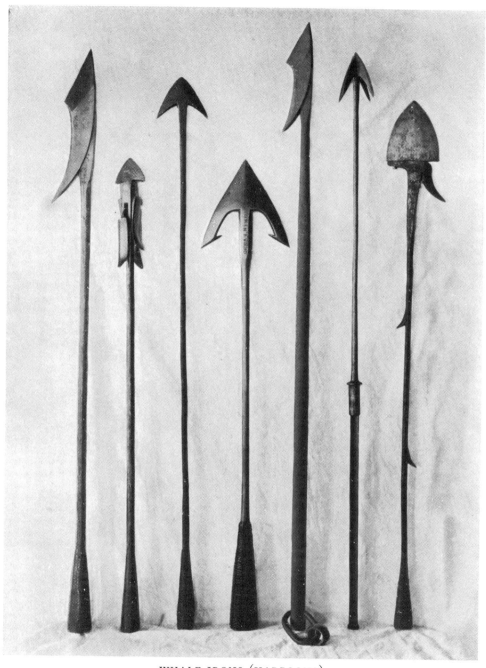

WHALE IRONS (HARPOONS)
Humpback iron (for recovering sunken whales), "lily" iron, long-shanked Arctic two-flued iron, an English iron with "stop-withers," Greener gun iron, Oliver Allen gun iron (patented 1845), Charles Freeman iron with explosive head (patented 1872)
In the collection of Mr. F. Gilbert Hinsdale

BOAT GEAR AND CRAFT
From left to right: Waif, lantern-keg, tub oar crotch, rowlock, shoulder bomb gun, wig-wag, paddle, boat-spade, darting-gun, toggle iron, lance. (A description of these articles may be found in the glossary.)
In the collection of the New Bedford Whaling Museum

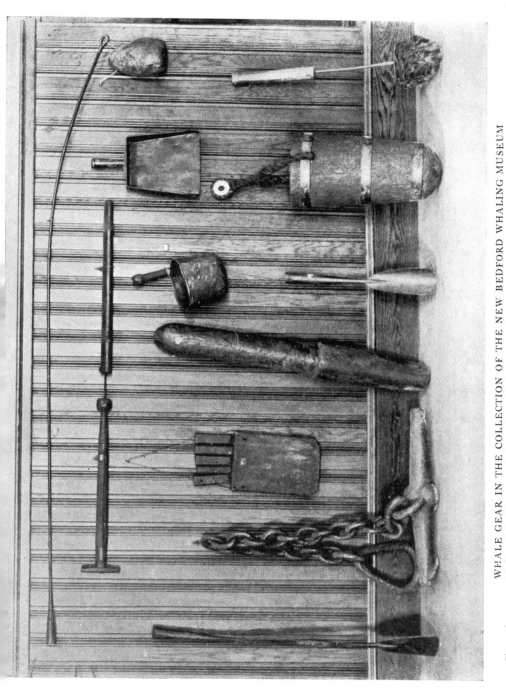

WHALE GEAR IN THE COLLECTION OF THE NEW BEDFORD WHALING MUSEUM
Top to bottom and left to right: Head-needle, boarding-knife, bone-spade, jaw-strap, blubber-knives, blubber-toggle, dipper, gouge spade, save-all, case-bucket, mast-head torch (for signaling boats at night)

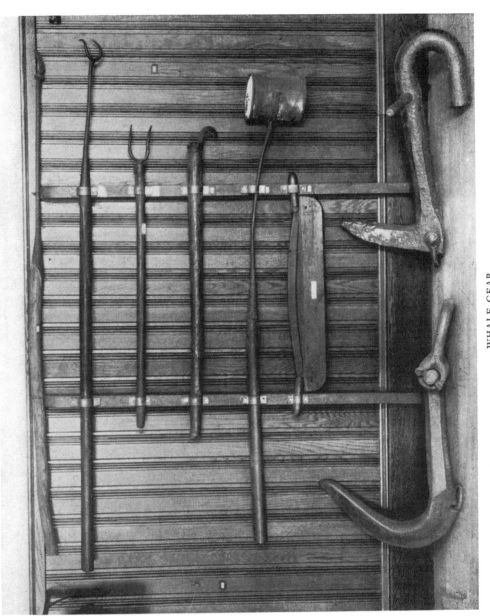

WHALE GEAR
Blubber-pike, two blubber-forks, blubber-gaff, bailer, mincing-knife (in sheath), blubber-hook, and ice-anchor
In the collection of the New Bedford Whaling Museum

CAULKER'S TOOLS AND RIGGER'S BELT
In the collection of the New Bedford Whaling Museum

GREENER BOW GUN
In the collection of the New Bedford Whaling Museum

WEIGHING WHALEBONE

CHAPTER XII
THE LAST DAYS OF WHALING

THE bark *Wanderer* was the last whaler ever to sail from New Bedford. She was wrecked outside the harbor in August, 1924. On the return of the schooner *John R. Manta* in 1925, the final whaling voyage was settled, and that spring the *Charles W. Morgan*, last ship of the old fleet, was towed to Round Hills, where she was beached and converted into a monument to a dead industry.

Along with his ships has passed the Yankee whaleman, and with him has passed the knowledge of his craft.

So late as the early nineties the Whaling Fleet of New Bedford numbered one hundred sail. Dismantled weather-beaten ships were berthed three deep along the wharf-sides. With fretted rigging and housed spars they lay for years awaiting a revival of whaling that never was to come.

For the fortunate youth of that day the unpoliced ships and grass-grown wharves made a marvelous playground. We learned to swim from the bob-stays of the old hulks. We contrived to paddle and row on rafts fashioned of hatch-covers, and used in boarding-parties overside. We swarmed over the rigging and slid down the backstays, spun the wheels, and on rainy days gathered in the cabins and played games and pretended one thing and another; and always it was something that smacked of the sea. Often in winter we skated up the river to where an old whaler was stranded, and there built fires of sheathing ripped from her sides. In those days we didn't "shoot" marbles; we "pooned" them with an overhand toss, just as a harponeer darts his iron; and when one boy was chased by another, he "went fluking" down the street. There were whole wharves along New Bedford's waterfront that served no purpose save that of a playground, and there was never another like it.

But earlier even than that, before my school days, I recall my uncle setting forth to join his ship at "Frisco." The Atlantic had been fished out, and the North Pacific and Arctic were the great whaling grounds. New Bedford ships were outfitting from the West Coast. I was five years old at the time, and I hid in my uncle's sea chest hoping to make the voyage with him. But I was discovered and his ship sailed without me, and I never saw my uncle again.

One of the sights of my childhood was the "boat train" which drove down the County road before sailing day, from the head of the river or from Acushnet; for many of the boat-builders were from the back country. A driver sat on the bow thwart of the forward whaleboat, maybe silk-hatted, always leather-booted. One horse only was required to draw each boat which rested on a skeleton cradle built on a long gear, with incurved arms at the after end, holding the boat on an even keel. A bucket dangled underneath the perch with which to water the horse. A flotilla consisted of six, seven, or even eight boats, and the succeeding ones were generally driven by farmboys who volunteered in order to see the harbor sights of the town.

In a vague way I gathered that petroleum had killed whaling; that the ships I knew were but the remnant of a fleet that had been; that the occasional sailing of a ship was a mere echo of stirring times when often a dozen cleared in a single day.

Acres of full oil-casks, buried in seaweed to keep them from shrinking in the sun's heat and chilling in the winter's cold, covered the old wharves, and along the alleyways between the casks the grass grew deep. Wrinkly-eyed old men paced the wharves gazing seaward, and every day was hushed as a Sabbath morning, for the business of the town had moved away from the harbor-front. Realization that whaling had received its deathblow came slowly to the old merchants and mariners of New Bedford. One of the most matter-of-fact of owners towed his favorite ship up the river and scuttled her in the berth where she had been launched fifty years before. Another vowed his fifty thousand gallons of oil would never be sold "till sparm went back to a dollar." In the years that were to follow, the seaweed blew away unheeded, the hoops rusted, the staves dried out, and his oil trickled and seeped down into the earth of the wharf unnoticed.

The Fourth-of-July whaleboat races were discontinued; people no longer turned out to see them, and oarsmen were becoming hard to find. Cotton manufacture was the transcendent topic of business discussion. Old men still gathered in the back shops of the two or three shipping offices that continued to carry on a precarious business, but they talked about ships and voyages of the past. The "Whaleman's Shipping List and Merchants' Transcript" ceased publication. One holiday a whaler was warped out into the stream and burned at night as a spectacle for the townspeople. It was a glorious sight — but it was a silent crowd that watched the dying embers.

Then the bottom dropped out of the whalebone market; horsewhips were no longer needed, and steel had displaced bone in umbrellas and stays. The few remaining ships on the West Coast sailed home to New Bedford.

At last the junking and scrapping of the fleet began. "Copper fastened and built on Honour," a rich field was opened; fortunes were made in junk.

The alleys and bystreets of New Bedford no longer echoed to the ring of the cooper's hammer on the resounding cask; the cresset had flickered out in the last cooperage. The last whalecraft shop and shipsmithy had closed its doors. There was no longer a single sparyard. One sailloft, one riggingloft, and one ropewalk were all that remained in what once had been the fifth seaport of America. Block and pump shops, boatyards and chandleries, all were empty or gone. The tap of the caulker's beetle was stilled forever. The old handicrafts were forgotten.

For years the death toll of the old shore craftsmen had kept even pace with the decline of the fishery. There were no apprentices, for it was clear the trades were dying. Along the wharves riggers and carpenters, caulkers and coopers, painters and shipsmiths worked side by side; and it was rare to see a man under sixty years of age, and few there were under seventy. Curious it was to see whitebearded patriarchs crawling over spars a hundred feet aloft, bending sail and serving rigging: curious and sad.

The ranks of the whitehaired old shipmasters that only yesterday paced the wharves have now thinned and passed on. There had been a fine light in the eyes of these old men and there was not a bowed head among them. Life to its very end they had found worth living; it may even have meant something to them that the world today has missed. For in this machinemade age we now live in, soft living has become the common ideal, we have killed adventure with quickened transportation, and the pioneer spirit is dying.

New Bedford relinquished her birthright too easily. From being accounted one of the most beautiful, and the richest per capita city in all the world, she stands today with many of her residential streets still elmshaded, but her harbor empty and three quarters of her outlying settled area a mass of wooden tenement houses, boasting the largest proportion of foreignborn population in any city of America. For three miles along her harborfront brick and cement factory chimneys belch forth blackened smoke where once tall masts and white sails raked the skyline. This was called progress. But the fragments that remain to tell the story of her whaling days are pre served in a museum the like of which is not to be found elsewhere along the seaboard. So replete with material was the old town that the mere backwash of her riches was sufficient to fill a museum to its doors. Stuff brought from all the Seven Seas, stuff fashioned by the whaleman himself, the accumulation of his leisure hours, and the actual tools of his craft, were fetched from the homes and storehouses of a hundred whaling masters and ship-owners and given to this institution to be preserved.

Among these relics of shop and wharf, of ship and distant sea, it may still be possible for another generation to evoke an image of the Golden Age of Whaling.

The record of the whaleman may be found in the files of the "Morning Mercury" and the "Whaleman's Shipping List." His own story written in his log-books has been in great part destroyed. One volume in fifty, perhaps, has escaped the rubbish heap. From these scant documents various histories have been compiled, and recent whaling fiction has been bolstered with a skeleton of fact.

New Bedford is still a wealthy city, with better schools, streets, public buildings, and utilities than most communities of her size; but she is no longer a great seaport. I am glad I lived in time to see the remnant of the old order, and there is little the world could offer that I would exchange for my memories of the ships, the sailors, and the old waterfront in this backwater of a bygone age.

Some day a cleaner craft than the Yankee Whaleboat may be evolved; some day man may fashion a machine more beautiful than a full-rigged ship; but I doubt if ever there will be a braver or a sturdier race of men bred in this world than the officers of that vanished fleet. There is one other thing that I hold certain: if ever there is to be fairer and better hunting than the chase of the Sperm Whale, man will have to voyage to other worlds to find it.

EPILOGUE

IT is twelve years since 'The Yankee Whaler' was first published, and during these years the Bowhead Whale of the Arctic and the Sperm Whale of the tropics have been practically undisturbed. Only the Eskimo and the shore whalemen of the West Indies and the Western Islands have continued the hunt which the modern Steam Whaler is not yet prepared to undertake. In this long period of security these whales have increased amazingly and it is inevitable that the steam whalemen will soon cast covetous eyes toward these unexploited and greatly augmented herds. Already depletion is evident among the Finback and Sulphur Bottom Whales which they have been hunting and various governments have been passing laws aimed at their conservation.

It will be interesting to see just what technique will be developed to make the new fisheries profitable. For new technique will be required. Old Yankee methods are forgotten and the present Steam methods will have to be revised to fit very different conditions. The Sperm Whale is a nomad and is not sought at present only because he is too seldom met with to pay for the coal consumed in cruising for him. Even when he is present he may escape detection as he stays only a short while at the surface and often remains below for the greater part of an hour.

Yet I imagine the Sperm Whale will prove the easier quarry of the two, for he can be captured by present methods if only he can be found in sufficient numbers. It would seem that an airplane lookout with telephonic communication with a mother ship would serve the purpose. A little altitude would widen the horizon and whales below the surface would also become visible.

It is the Arctic fishery that presents the more baffling problem. New England whalemen approached the Bowhead Whale under sail or paddle. The sound of oars was sufficient to gallie him. Once among the ice he could not be followed, and ice is never far distant in the Arctic.

All the more obvious weapons have been experimented with against whales: harpoons loaded with acid have been tried, swivel guns and shoulder guns have been in actual use for years. Perhaps the most effective explosive weapon ever invented was a "darting gun," which was thrown like any other harpoon and carried a bomb as well as a whaleline. Japanese whalemen have employed steel nets in catching Finbacks and these might prove successful against the Bowhead. The Simple Simon method, so far as I know, has never been tried out; apparently the problem of bait bulks too large. Mines, depth bombs, and projectiles dropped from the air are still untried and a controlled torpedo bearing a whale line is among the possibilities. But it would seem that success in capturing this whale will probably hinge on the development of a silently driven boat in which to carry the harpoon gun. Only one thing may be prophesied with certainty, which is that at no remote date these two whales will again become commercially important.

Clifford W. Ashley

Westport, Mass.
March 23, 1938

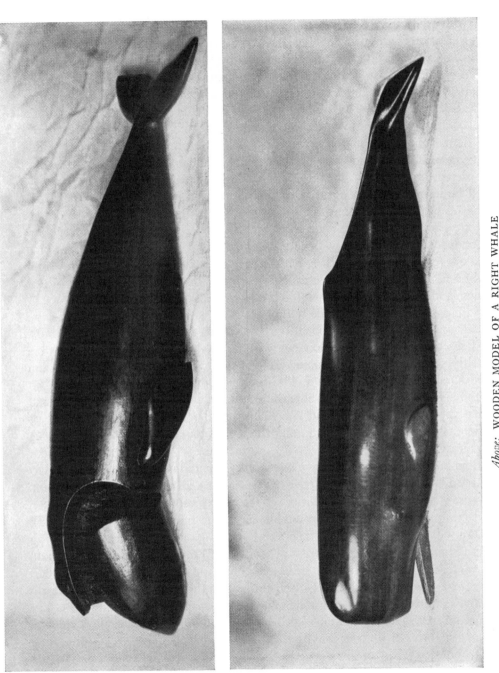

Above: WOODEN MODEL OF A RIGHT WHALE
In the collection of the New Bedford Whaling Museum

Below: SANDALWOOD MODEL OF A SPERM WHALE (MADE BY CAPTAIN ALBERT ROBBINS)
In the author's collection

Right: A SIXTEEN-FOOT SPERM WHALE'S JAW
The crosspiece is a two-foot rule
In the collection of the New Bedford Whaling Museum

Left: WHALEBONE: BALEEN FROM THE MOUTH OF THE BOWHEAD WHALE

These two photographs are in scale and show the length of bone and jaw compared with the height of a man

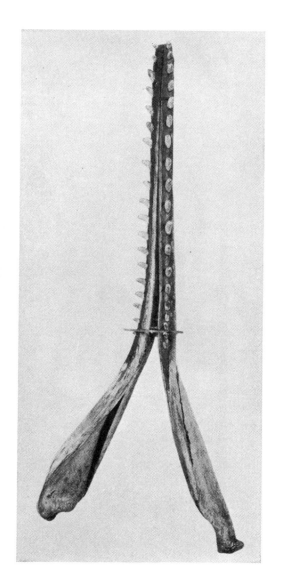

FLUKES OF A SEVENTY-FIVE BARREL SPERM WHALE
Photograph by the Whaling Film Corporation

WHALING COSTUMES FROM THE SUNBEAM, 1904

The trousers at the right were worn by the mate. The foot-gear is typical and constituted a part of the author's outfit. All are wooden-pegged and in the slop-list are described (left to right) as brogans, boots, and pumps. The boat crews went barefooted

THE CAPTAINS OF THE "STONE FLEET"

Left to right, standing: Capt. Beard, Capt. Gifford, Capt. Swift, Capt. Childs, Capt. Stall, Capt. French, Capt. Wood, Capt. Cumiski, Capt. Willis, Capt. Bailey. Sitting: Capt. Molloy, Capt. Swift, Capt. Brown, Capt. Howland, Capt. Worth, Capt. Tilton, Capt. Brayton, Capt. Taylor, Capt. Chadwick

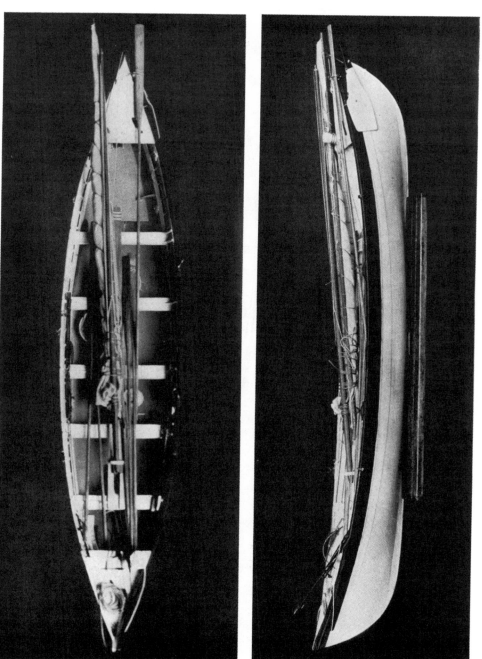

MODEL OF A WHALEBOAT
In the author's collection

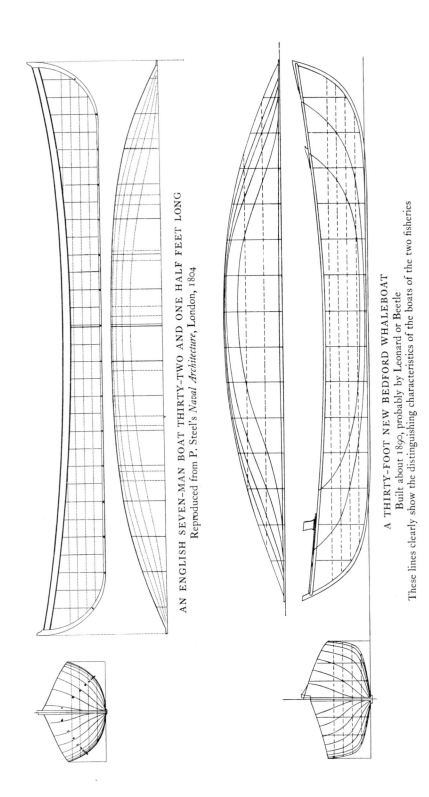

MODEL OF THE LAGODA
In the collection of Mr. Frank Wood

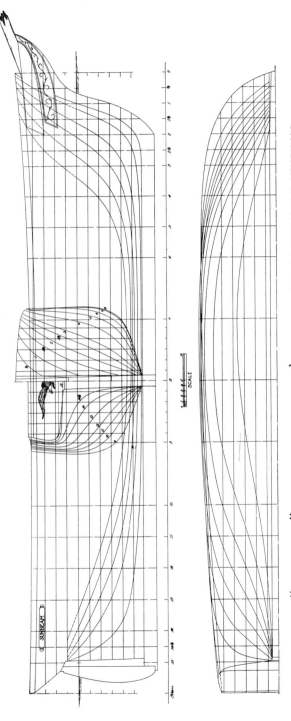

LINES "TAKEN OFF", THE ORIGINAL BUILDER'S BLOCK MODEL OF THE BARK SUNBEAM
255.31 tons, length 106.4 feet, beam 27.3 feet, depth 15.2 feet. Built for J. and W. R. Wing at Mattapoisett in 1856 and called "The Whaling Yacht Sunbeam"
In the collection of the New Bedford Whaling Museum

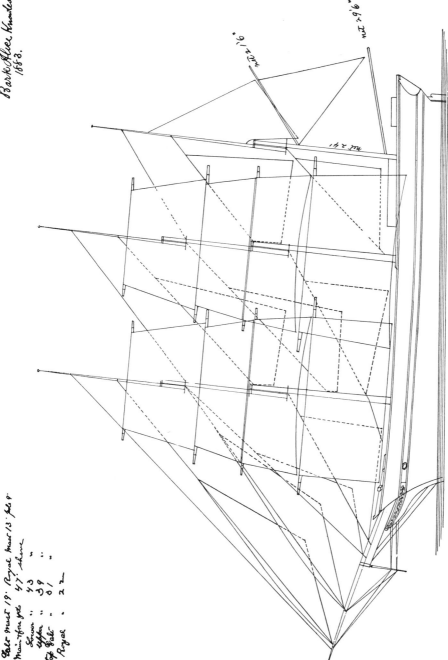

SAIL-PLAN MADE IN HART'S SAIL LOFT, NEW BEDFORD, IN 1883, FOR THE BARK ALICE KNOWLES

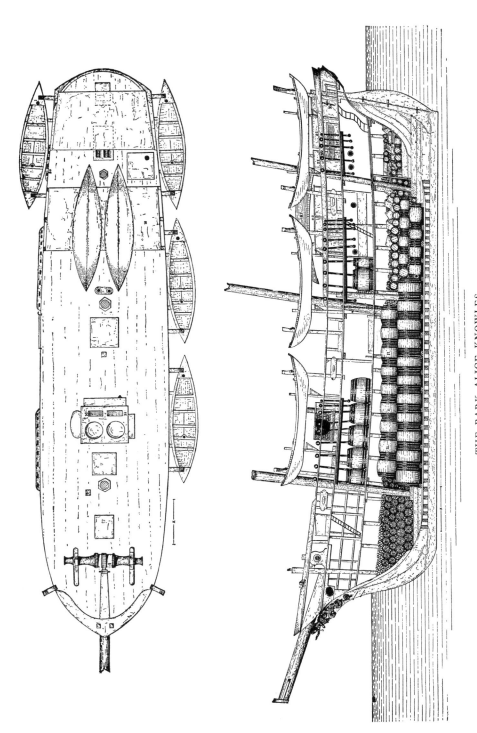

THE BARK ALICE KNOWLES

302.78 tons, length 115 feet, beam 28 feet, depth 16.7 feet, built at Weymouth, Mass., 1878, for J. P. Knowles, 2nd. Reproduced from *The Fishery Industries of the United States*, Washington, 1889

A GLOSSARY OF WHALING TERMS
TO WHICH ARE ADDED CERTAIN WORDS APPLYING TO THE TRADES OF COOPER, CAULKER, AND RIGGER

ADVANCE: Money allowed to whalemen before starting on a voyage, for the purchase of outfits and for settlements with the boarding-master. It is charged against their subsequent earnings.

ADZE, COOPER'S: A small short-handled adze, similar to a brickmason's hammer.

AFTER HOUSE: The whaler's "Round house."

AFTER OAR: The stroke oar of the whaleboat. He sits on the port side of his thwart, and his rowlock is on the starboard gunwale.

AGENT: The managing owner of a whaleship.

AH! BLOWS! See "There she blows!"

AIR UP, TO: To blow up stronger, said of the wind.

ALL GONE, SIR! The sailor's reply after obeying the order "Cast off!" or "Let Go!"

ALOW FROM ALOFT! The lookout is called to deck with this order.

ALOW: The deck, as distinguished from Aloft. The hail from the masthead to the deck is "Alow, there!" There is little doubt that the landsman's hail, "Halloa!" is a corruption.

AMBERGRIS: A foreign substance found in the alimentary canal of a Sperm Whale, used as an agent in perfumes.

APEAK: The position of the oars when the boat is fast to a whale.

ARTICLES: The ship's papers. Signed by all hands when shipping.

AWAY: Said of the boats when lowered, as, "They lowered away"; "Three boats were away." More stress is put on the second syllable than is customarily employed.

BAILER: A long-handled copper dipper, holding about two gallons. For removing oil from try-pots.

BAILING: Process of removing spermaceti from the case, or head of the Sperm Whale.

BAILING-BUCKET: See Case-Bucket.

BALEEN: Black whalebone from the mouths of toothless whales.

BALEEN WHALE: All whales, except Sperm, that were commercially hunted.

BARREL: A barrel exists aboard a whaleship only as a unit of measure for oil, $31\frac{1}{2}$ gallons. Everything is carried in large casks.

BEARDS: also called stop-withers. Small reverse barbs on an English two-flued harpoon. See illustration.

BEARERS: Upright stanchions between the davits upon which the boat cranes are hinged.

BECKETS: Chest-handles of rope, often elaborately knotted.

BEETLE: A caulker's heavy driving-mallet.

BELLOW: The sound made by the Right Whale when in violent action. Whales have no vocal cords. The sound may be in the lungs or it may be abdominal.

BESET: Progress stopped by the closing-in of ice.

BIBLES, *or* BIBLE LEAVES: See Books.

BILGE-HOLE: Bung-hole.
BILGE OF A CASK: The fat or bulging part, the waist.
BISCAY WHALE: North Atlantic Right Whale.
BLACKFISH: A small cetacean. The forehead or "melon" makes the finest lubricant known. Used exclusively for watch oil.
BLACKSKIN: The thin slimy outer covering of a whale, so tender that it is easily scraped off with the finger-nail.
BLACK WHALE: An inclusive term for all commercially hunted whales except Sperm and Fin whales; that is, Bowhead, Right, and Humpback.
BLANKET, or BLANKET-PIECE: A long strip of blubber hoisted from the whale to the maintop. Subsequently cut into smaller horse-pieces before being minced.
BLASTED WHALE: Swollen from the formation of gas in the belly. Also a decomposed whale.
BLINK: "Halation" over ice and show.
BLOW, A: A spout. The moist visible breath of a whale.
BLOWS! See "There she blows!"
BLOW, TO: To spout. The whale's act of breathing.
BLOW UP! The order given when the time has arrived to inflate the poke.
BLUBBER: Thick oily outer casing of the whale which serves as protection and insulation against pressure and cold.
BLUBBER-FORK: A long-handled fork for tossing blubber-"books" into the try-pots.
BLUBBER-GAFF: A short-handled hook for dragging blubber about deck.
BLUBBER-HOOK: An iron hook, from fifty to one hundred pounds in weight, suspended from the cutting-tackle, for hoisting blubber.
BLUBBER-HUNTER: A whaler.
BLUBBER-PIKE: A single-pronged instrument for pushing and forking blubber about deck.
BLUBBER-ROOM: The space in the upper hold near the main hatch reserved for the temporary storage of blubber, when deck space is insufficient. If a whale is small and the weather good, the blubber-room is not used.
BLUBBER-TOGGLE: An oaken pin buttoned to a strap in the blanket-piece, used instead of a blubber-hook.
BLUE WHALE: A Sulphur-Bottom.
BOARD HO! The warning shouted from the gangway when a blanket-piece is about to swing inboard.
BOARDING-KNIFE: The two-edged long-handled sword-like knife which severs the blanket-piece.
BOARDING-MASTER: A crimp who housed and supplied men for the crews of whalers.
BOAT-CREW: The six men who comprise her full complement, or the four men who row a whaleboat, generally the former.
BOAT-CREW WATCH: When on the whaling grounds instead of the starboard and port watches serving "watch and watch," boat-crew watches were established. As there were four boats, a quarter of the crew instead of one half was kept on deck. The men were fresher and more alert as a result.
BOAT-CROTCH: A forked upright in a whaleboat on the starboard gunwale forward,

A GLOSSARY OF WHALING TERMS

to hold the "live irons"; that is, the first and second irons. Called a "mik" in the British Greenland Fishery.

BOAT-DAVITS: The curved wooden arms that suspend a whaleboat overside.

BOAT-FALLS: The hauling ends of the davit tackles.

BOAT-GEAR: Everything in a whaleboat except the crew and the tools of capture; that is, not the guns, harpoons, lances, and spades.

BOAT-HEADER: The man who steers the boat in going on a whale, and afterwards kills it. Generally a mate, but sometimes an experienced whaleman with no ship duties save masthead and cutting-stage, whose only title is boat-header.

BOAT SKIDS: Stern davits for a spare boat. Sometimes called "tail feathers." Also an unroofed frame formerly used instead of a forward house.

BOAT-SPADE: A short-handled spade carried in a boat and used to cut a hole when attaching a line to a dead whale before towing. See Fluke-Spade.

BOAT-STEERER: Harponeer. The man who pulls the harpoon oar, darts the iron into the whale, and then steers while the mate or boat-header lances him.

BOILING: Trying-out.

BOLLARD: English name for loggerhead. Usually a bollard is an upright timber on a wharf.

BOLT, TO: Said of a whale when he half breaches and starts to run.

BOMB-GUN: Heavy shoulder gun which fires a bomb-lance.

BOMB-LANCE: Shot from a shoulder gun. Generally a brass tube about fourteen inches long filled with explosives and fitted with a short time fuse. It is iron-pointed and "feathered" at the other end.

BONE: See Whalebone.

BONE, TO (an iron or lance): To strike a rib or other bone with the implement. This generally implies a miss.

BONE-SPADE: A spade with a long flat shank.

BONNET: Cheever defines the bonnet of a Right Whale as being "the crest or comb where there burrow legions of barnacles and crabs, like rabbits in a warren, or insects in the shaggy bark of an old tree." It is a pitted horny cap situated on the nib end of the Right Whale's snout, and it is generally infested with whale-lice.

BOOKS: Bibles or minced horse-pieces. The blubber is sliced thin, and left adhering to a rind, so the pieces can be easily forked into the pots.

BOOTS: All whalemen's footgear is of leather, wooden-pegged to prevent wear on decks. Oil rots rubber.

BOW BOAT: The boat on the forward davits on the larboard side.

BOWHEAD WHALE: The Arctic or Polar Whale, same as Greenland Whale. He gives the greatest amount of oil and bone.

BOW OAR: The second oar next to the harponeer. This is the third responsible position in a whaleboat. The bow oar sees that all whalecraft is clear, and attends the wants first of the boat-steerer and then of the mate. In pulling up to the whale, for lancing, it is his duty to grasp the line forward of the chocks and fetch it to the bow. This causes the boat to "veer" and tow parallel to the whale. His task requires both strength and a horny palm.

Box: Sunken cuddy-board in the bows of a whaleboat, where the forward end of the whale-line is coiled.
Box-Line: The line coiled in the bow box.
Box-Warp: Same as Box-Line.
Breach, a Full: A whale's leap, clear of the water.
Breach, a Half: A whale's leap, partly clear of the water.
Bread-Bag: Canvas bag used in the forecastle for holding hard bread.
Break Water, to: When a whale first comes to the surface he is said to break water.
Breaming: The method of cleaning a ship's bottom, formerly employed by whalers when heaving down. Torches either of twigs or oil-soaked oakum were held against the bottom. This loosened the weed, grass, slime, goose clams, and barnacles that encumbered the sheathing, so that it was easily scraped and brushed off.
Bring-to, to: To come to a full stop, said of whales.
Brit: Small red shrimp-like crustaceans that float in fields or beds near the surface of the sea, the food of the Baleen Whale.
Brogans: A high Blucher shoe with only two eyelet holes at each side, to admit of easy removal.
Broken Voyage: An unprofitable voyage.
Bull: The male whale.
Bunch: The prominence at the back of a sperm whale's neck.
Bung-Borer: A two-handled auger for making a bung-hole.
Bung-Hole: The hole in the side of a cask. Formerly called bilge-hole.
Bung-Starter: A wooden-headed hammer with a thin flexible hickory handle. The bung is loosened by tapping around the hole.
Bung up and Bilge Free: The best position for a cask.
Buried to the Hitches: A successful dart. The iron has penetrated to its socket. See Hitches.
Busk: A wide front stay for a corset. Frequently made at sea of white bone.
Buy in or Own in, to: To acquire or have a share in a whaleship.
Cachalot: French name for Sperm Whale.
Calf: Sucker, the young of the whale.
California Grey: A humpless whale with two "ginger rolls," often mistaken for a Right whale.
Camboose: Whaler's name for caboose or galley.
Can-Hooks: Iron hooks used in hoisting casks.
Cant-Hook: Used in spar-yards to roll over spars. The same as cant-log, and similar to peavy, except that the pole of the latter is iron-shod.
Case: The forehead of a sperm whale. It is outside the skull, and is composed almost entirely of spermaceti. See Junk.
Case-Bucket: For bailing the spermaceti from the case. It has a round or pointed base, and is forced down into the case with a beam and tackle.
Cask: General name for the large barrels employed on a whaler.
Caulk, to: To drive oakum into a ship's seams to make them tight.
Caulker: A man whose trade it is to caulk.

A GLOSSARY OF WHALING TERMS

CAULKER'S SEAT: A small wooden box which also contains his tools.

CAULKING-IRONS: Short iron chisels with which oakum is directed into the seams of a ship. See Horsing, Making, Rasing, and Reeming irons.

CAULKING-MALLET: A short-handled, long-headed mallet of wood, iron-hooped. See Beetle, Reeming-Beetle.

CHAMFER KNIFE: A heavy iron-handled draw-knife with which to bevel the inside of the chines or stave-ends.

CHAWED BOAT: A boat which has been chewed by a Sperm Whale.

CHIMNEY: "His chimney's afire!" was the Amagansett exclamation when the whale began to spout blood.

CHINCING IRON: A caulking-iron used by a cooper to caulk around the head of a cask.

CHINES: The rim of a cask.

CHINE-HOOKS: Same as Can-Hooks.

CHINE-TO-CHINE: Casks stowed end-to-end.

CHIPS: On merchant ships, the carpenter; on whalers generally the cooper.

CHOCK-PIN: A slender oak pin passing through the chocks to keep the whaleline from jumping. Sometimes made of black whalebone.

CHOCKS: A groove in the stem of a whaleboat through which the whaleline leads. It is either bushed with lead or fitted with a bronze roller.

CHURN, TO: The boat-header churns his lance when he works it up and down in the whale's anatomy without withdrawing it.

    A whale's flukes are said to churn when their up-and-down motion makes a commotion on the surface.

CLEAN SHIP: A whaler without oil. In Starbuck (p. 190) is the following entry regarding the sloop *Keziah* of New Bedford: "Lost a man overboard, and returned home clean."

CLEAR AWAY, TO: To prepare for lowering. Gripes are cast off, line-tubs are put aboard, the boat-falls are cleared, and stops are removed from whalecraft and sails, preparatory to the order, "Lower away!" See Clear Boats.

CLEAR BOATS, TO: To give them a preliminary overhauling when whales are first sighted. Lines of drying clothes are removed from between the davits, stowed articles are taken out, the boat-plug is tapped home, line-tubs are uncovered, and the water-keg is examined. See Clear Away.

CLUMSY CLEAT: A thick thwartship plank which forms the after edge of the bowbox. The edge is notched to fit the officer's left thigh, and so steady him at his job of darting and lancing.

COMMANDER: A heavy wooden rigger's mallet.

COOLER: A metal tank in which oil stands after boiling. There is a small one beside the try-works and usually two large ones between-decks.

COOLING DOWN: Said of a whaler when her fires are dying or out, and the last gallon of oil has been bailed to the coolers and is waiting a temperature low enough to permit "drawing off."

COOPER'S ADZE: See Adze.

COOPER'S DEVIL: His portable anvil. It usually stands on a cask-like pedestal, and it has no horn.

Cooper's Vice: A small implement resembling a corkscrew used to lift the head of a cask.

Cow: Female whale. Scammon, in his "Marine Mammalia," describes the cow Sperm Whale as about one third or one fourth the size of the male. "She is likewise more slender in form and has an effeminate appearance."

Crab: A winch used in the early days of Nantucket to haul a whale out on the beach. The power was insufficient to haul him to dry land. It merely got the whale into shoal water at high tide, and he was cut-in as the tide receded. The carcass easily floated off at the next high tide, since its displacement was considerably reduced.

Craft: See Whalecraft.

Cranberry Pudding Voyage: An early name for Plum Pudding Voyage, which see.

Cranes: Hinged triangular brackets of wood which swing out from the bearers and support the boats.

Cresset: An iron, basket-shaped grate, filled with burning chips and employed to heat casks when coopering; also a deck torch used when cutting and boiling at night.

Crossjack Yard: Pronounced "Crotchet" by whalemen; "Crojek" by merchantmen. The name of the lower yard on the mizzenmast.

On a whaling bark this sets no sail, and consequently is called "dummy crossjack." The mainbraces lead to it.

Crotch: See Boat Crotch and Tub-Oar Crotch.

Crow's-Nest: An enclosed shelter for the lookout, used only in the Bowhead Fishery.

Croze: (1) A plane for grooving the end of the staves. It makes a slot which holds the head of a cask.

(2) The groove into which the head of the cask is fitted.

Cruise: A whaler was said to "cruise" when, on the whaling grounds, she either lay-to or went under shortened sail at night, and set sail again in the morning. She tacked back and forth across the grounds, heading westward in the mornings if possible, and eastward in the afternoons, so that a spout ahead could be seen in a favorable light.

Cuddy Board: A short deck over the bow or stern of a whaleboat.

Cuntlines: Spaces between the sides of stowed casks.

Cut, a: Any number of dead whales alongside a ship at one time. "We had five whales in a single cut."

Cut Flukes, to: To lift the flukes out of the water and to strike out with them.

Cut-In, to: The process of removing blubber from a whale. See "flenzing," and "flinching."

Cutting-Gear: Apparatus used for cutting-in a whale.

Cutting-Spade: A wide, flat, long-handled chisel-shaped implement for cutting blubber. Specifically, one used on the cutting-stage, as distinguished from one used on deck which is called a deck spade.

Cutting-Tackle: Huge blocks and falls hung under the maintop and led to the windlass, which hoist the blubber over the main hatch when cutting-in.

A GLOSSARY OF WHALING TERMS

DART, TO: The harpoon is darted, hove, pitched, or tossed — but *never* thrown.

DARTING DISTANCE: Near enough to the whale to strike. The limit is about thirty feet. Fifteen feet is considered "good darting distance," and "wood to black-skin" (actual contact with the whale) is preferred in the Sperm Fishery.

DARTING-GUN: A gun carrying both a bomb and a harpoon. The whole apparatus is darted. (Illustrated, and also described in the text.)

DARTING-IRON: The peculiar harpoon of the darting-gun; also the whole instrument.

DECK-POT: A try-pot not enclosed in the brickwork of the ovens and used either to hold scrap or cool oil.

DECK-SPADE: A short-handled cutting-spade for use on deck.

DEVIL FISH: A whaleman's name for the California Grey Whale.

DITTY-BOX AND BAG: The first is shaped like a wooden spice-box and generally holds a sailor's toilet articles, brush, comb, and mirror. The ditty bag is about a foot long, and contains needles, thread, buttons, knives, etc.

DOCTOR, THE: The cook of a whaler, a much-maligned man.

DONKEY'S BREAKFAST: The corn-husk or straw mattress of the forecastle. The outfitter's price used to be a dollar apiece, and one was part of each sailor's outfit.

DOUBLE CARD STEERING COMPASS: A "transparent compass" with a card visible either from above or below. It hung inside the after coaming of the companion or cabin skylight, and served both as a steering compass and a telltale.

DOUBLE-ENDER: A boat with a sharpened stern similar to the bow. A canoe is a double-ender, and so were all recent whaleboats except some in foreign fisheries that were made square-sterned to support heavy bow guns.

DRAW-OFF, TO: To fill casks with oil from the coolers.

DRAW, TO: Said of a harpoon when it pulls out. "We drawed our iron." "Both our irons drew."

DRIFT: The set of a current either with or without ice.

DRIFT ICE: Broken sheet ice.

DRUG, DROGUE, DRUDGE: A square block of wood fastened to a whaleline and used to check the whale. Nowadays only used when the whale is about to take the last of the line. It acts against the water exactly as a kite does against the wind. Derived from "drudge," the earliest form of the word ("to labor patiently for others"), and not from "drag," as usually stated.

DUMMY CROSSJACK: See Crossjack.

DUNGAREES: Overalls and jumpers.

DUNNAGE: Wood for chocking, used in stowing casks and other cargo. About the same as cord wood. The same pieces carried voyage after voyage became smooth, round and oil-soaked.

ENTER, TO: To penetrate, said of a whale iron, as, "His iron entered."

EYE TO EYE, FROM: The scope of a Right Whale's flukes when "sweeping."

FALL!, A: A strike. A term used in the British Greenland Fishery, meaning that a boat had got fast.

FAST: Signifies that the boat has harpooned a whale and is attached by means of the line.

FAST BOAT: A boat fast to a live whale.

FAST FISH: A live whale with a boat in tow.
FASTEN, TO, *or* TO GET FAST: To strike or harpoon a whale.
FENKS: Name for scrap in British Greenland Fishery.
FID: A large, long-pointed wooden or bone tool for splicing, similar to a marline-spike, except the latter is of iron. See Pricker.
FIDDLES: Racks which fit on top of the cabin table. They serve to keep dishes in place in a sea-way.
FIGHTS AT BOTH ENDS: Said of a Sperm Whale which fights with both jaw and flukes.
FIGHT SHY, TO: Whales are said to "fight shy" when they will not allow boats to approach.
FINBACK: The commonest of whales. A medium-sized variety with high dorsal fin and very racy lines.
FIN-CHAIN: A short chain strapped around the small of the fin. Used for the first blanket when cutting-in a Right Whale.
FINNER WHALE: See Fin Whale.
FINNING: The action of a whale when, listing to one side, he strikes and splashes the water with his fin.
FIN OUT OR FIN UP: The position of a dead whale, as, "He is fin out"; "He turned fin up."
FINS: The flippers of a whale.
FINS: English name for whalebone, which see.
FIN WHALE: A term which includes the Finback and Sulphur-Bottom, both of which have prominent dorsal fins.
FIRE-PIKE: An instrument for poking the try-works fires.
FISH: General name for whales, used by whalemen.
FISHERY: A name applied to the entire fleet of any town, nation, or locality. Also applied collectively to the ships fishing on an important whaling ground; that is, the New Bedford Fishery, the American Fishery, the Sperm Fishery, the Arctic Fishery.
FIVE AND FORTY MORE! Shouted by the crew when the last piece of blubber from any catch is swung inboard. It refers to forty-five barrels, which is the average take from a single whale. No matter what the size of the whale may be, the call is never varied.
FIVE-BOAT SHIP: One that lowers five boats.
FLAG: "There's the Flag!" or "The Red Flag's Flying" means that the whale is spouting blood.
FLAG A CASK, TO: To put rushes between the staves, a method of caulking.
FLAGGING-IRON: An iron instrument with which to pry apart two staves of a cask so that a flag or rush may be slipped between them.
FLEMISH COIL: A layer of line in a tub. The line is coiled spirally layer upon layer.
FLENZE, TO: To cut in, the old English term.
FLESH AN IRON, TO: To drive it through the blubber and into the meat.
FLINCH, TO: To cut in, the modern English term. Said to have been an old Nantucket term.

A GLOSSARY OF WHALING TERMS

FLIPPER: The fin of a whale, or the hand of a sailor.
FLOE: A large sheet of ice.
FLUE: The barb of a whale harpoon, probably from "fluke."
FLUKES: The horizontal tail of a whale.
FLUKE-CHAIN: The chain strap put around the "small" (the root of the tail) which holds a whale alongside. It leads to a hawse-pipe forward.
FLUKE-SPADE: A name applied to the boat-spade when it is used as an offensive weapon for "hamstringing" a whale. It carries a light line by which it is recovered.
FLUKES, TO CUT: To strike out with the flukes.
FLUKING: The gallied flight of the Right Whale, in which his flukes are lashed out of the water from side to side.
FLURRY: The dying struggle or flight of a whale. As a whale weakens he lists to one side and swims in a narrowing circle making a last effort to escape. At the end he summons his reserves and frequently dies in the midst of a terrific commotion of thrashing flukes.
FOREGANGER: The English iron had an eight- or nine-yard warp hitched to it. This was kept coiled and stopped to the end of the pole until going into action, when the end was loosened and bent to the whaleline.
FOREHEAD: The front elevation of the Sperm Whale's head.
FORE-AND-AFT CUTTING-STAGE: Short single-plank stagings, hung overside.
FORWARD HOUSE: A roofed superstructure which supports the spare boats.
FOOTGEAR: All whalemen's footgear was made to admit of instant removal. In the Sperm Fishery the men lowered either barefooted or in stocking feet. In the Arctic Fishery they wore either pumps or brogans which could be kicked off instantly. Sperm whalemen frequently went barefooted in warm weather, but this was not permitted when cutting-in, as toes were liable to be clipped. Leather boots were worn at this time, and also in cold or wet weather.
FOUL LINE: A whaleline which has kinked or looped, and caught hold of some object in the boat.
FOUR-BOAT SHIP: One that lowers four boats.
FRITTERS: Early English name for scrap.
FU-FU: Mush and molasses.
FULL SHIP: A ship with a full cargo of oil, hence, homeward bound.
GAFF: See Blubber Gaff.
GALLIE or GALLY, TO: To startle or frighten a whale.
GALLIED: Frightened, said of a whale.
GALLOWS: A name sometimes applied to the skid beams.
GAM, A: A visit between whaleships at sea, elaborated upon in the text.
GAMMING: Visiting, as practiced between whaleships at sea, also said of Sperm Whales when they are herded and not in motion.
GEAR, BOAT: Material other than craft carried in a whaleboat.
GET FAST, TO: See Fasten.
GIG TACKLES: Small fore-and-aft tackles to secure whaleboats on the cranes.
GINGER ROLLS: The whaleman's name for the belly furrows on Finback, Sulphur-Bottom, California Grey, and Humpback whales.

"Give it to him!" The invariable order from the boat-header to the harpooneer, directing him to dart his iron.
"Giving the Whale the Boat": To tie the end of the whaleline to the boat and then jump overboard, a practice of the British Greenland Fishery.
Glip: English name for slick, which see.
Go Down, to: To sound.
Going Quick: A common log-book entry is "Raised whales, going quick."
Go into his Flurry, to: To commence the death struggle, said of a whale.
Go on the Whale, to: To row or sail up to him for the purpose of getting fast.
Goney: Whaleman's name for the albatross.
Goose-pen: The water tank under the tryworks.
Gouge-Spade: A half-round spade with which to cut holes in blubber, for reeving chains or ropes or embedding hooks.
Grains: A harpoon-like instrument with several barbs, something like a trident, used for fish, seals, etc.
Grampus: A small variety of cetacea often found close to ships. They grunt or puff very audibly; hence the expression "puffing like a grampus."
Grapnel: A five- or six-fluked grappling-hook carried in a whaleboat for recovering lost whaleline.
Great Rorqual: The Sulphur-Bottom Whale, also called Blue Whale.
Greener Gun: A heavy swivel bow gun for shooting harpoons. Used in the Scotch Fishery and in the California Bay Fishery the last half of the nineteenth century.
Greenie or Greenhorn: Inexperienced man on his first voyage whaling.
Greenland Whale: The Bowhead Whale of the Atlantic Arctic.
Gripes, Boat: The lashings which hold a boat laterally secure on the cranes, after hoisting.
Grounds: See Whaling Grounds.
Ground Tier: The bottom layer of casks in the lower hold, always stowed "fore-and-aft."
Gurry: The slime, oil, blackskin, etc., that encumbers the deck while cutting-in.
Hail from, to: (1) To come from. (2) To announce your home port.
Hand-Lance: A long-shanked instrument, flat-headed and sharp-edged, for killing a whale. It has a six-foot pole for a handle which fits into a socket, giving a total length of eleven or twelve feet.
Hamstring: To sever the fluke tendons at the small, with a boat-spade, in order to stop a running whale.
Hard Bread: Thick square biscuit, made very hard in order to resist dampness and deterioration.
Hardtack: Another name for the above.
Harness: All chain gear attached to a whale while cutting-in, jaw-strap, fluke-chain, etc.
Harness Cask: The cask in which brine is soaked from salt beef preparatory to cooking. The cask generally is partitioned, and the beef stays one week in each of the two sections.
Harping Iron: The old name for harpoon.

A GLOSSARY OF WHALING TERMS

HARPOON: An iron or steel instrument with a barbed head for fastening to whales. It is mounted on a pole and is commonly called an "iron."

HARPOON OAR: The forward oar of a whaleboat which is pulled by the harponeer in approaching a whale.

HARPONEER, *or* HARPONIER: The boat-steerer who pulls the forward oar in the boat, harpoons the whale, and then steers while the mate kills it.

HARPOONER: A term not used by whalemen. See Harponeer.

HEAD TO HEAD, TO GO ON: To attack a whale "head first" or bow on.

HEAD, TO: To command a whaleboat; that is, to fill the office of mate or boat-header.

HEAD MATTER: Spermaceti from the case of the sperm whale.

HEAD OUT: When a whale is said to be "head out" it implies that he is swimming rapidly, "going quick."

HEAD-SPADE: A long and round-shanked cutting-spade for separating the case from the junk and the junk from the white horse.

HEAD-STRAP: A chain strap for hoisting the case and junk.

HEAVE DOWN A SHIP, TO: To careen. Whalers being of very strong construction were "hove down" long after the practice had been given up in other services. The last vessel to "heave down" at a New Bedford wharf was the *Josephine* in 1893. It was a custom of the New Bedford ships in the South Seas to remove their cargoes at a convenient island, and with heavy tackles to heave their ships over on their beam ends in order to repair and clean bottoms.

HEAVER: An instrument with which to pull or tighten a strand. Used in knotting or splicing. A rigger's tool.

HEAVING-BAR, COOPER'S: An oak beam five or six feet long with a flange near one end, similar to a can-hook which gripped the chines of a cask. To the other end was attached a rope. A dozen or fifteen men tailed on to this rope, and turned over full casks of oil on the wharf. If a low horse was placed beside the cask, and the rope was hove on smartly, the cask was "tripped" over the horse and turned completely "end for end."

HEAVING-BEETLE: See Reeming-Beetle.

HEN FRIGATE: Any ship with a woman aboard. Not infrequently a captain took his wife with him on a voyage. Many a woman has wintered north of the Arctic Circle, and many a New Bedford child has been born there.

HERD, TO: Said of whales when they gather together on the feeding ground.

HITCHES, THE: The round-turn and eye-splice of the line at the socket of a harpoon.

HOLD, TO: To remain fast, said of an iron, as, "His iron held."

HOLIDAY, A: Any slighted task on shipboard, specifically an unslushed place left on a spar.

HOLLOW THE BACK, TO: A trick of the Right and Bowhead Whales. An iron will not penetrate the slack blubber so formed.

HOOK ON, TO: To get fast. "We hooked on to a whale."

HOOKER, OLD: Familiar name for a whaler.

HOOP OF A CASK: One of the encircling members that keep the staves of a cask together.

HOOP-DRIVER: A cooper's tool, a grooved instrument to be held against a hoop and struck upon with a hammer to drive the hoop into place.

HOOPS, THE: A pair of spectacle-shaped rings bolted to either side of the royal-mast breast high above the topgallant crosstrees. Their purpose was to steady the lookout.

HORSE-PIECES: Pieces of blubber cut from the blankets, about six inches wide and several feet long.

HORSING-IRON: An iron with a withe handle (to spare the hands) used by a caulker for heavy driving in a deep seam. Held by his helper.

HOSE-SCUTTLE: A scuttle in the deck through which the hose is led in filling casks that have been already stowed.

HOWEL: A cooper's plane for smoothing the inside edge of a cask.

HUMP: The dorsal fin of a Sperm or Humpback. (The Right, California Grey, and Bowhead Whales have neither humps nor fins on their backs.)

HUMPBACK WHALE: An inshore whale with very long curved flippers and a humplike dorsal fin.

HUMPBACK, TO: To fish for Humpbacks.

ICE ANCHOR: An ice anchor has but a single fluke, which is dropped into a hole cut in the ice. See illustration.

IRON: The name commonly applied by whalemen to a harpoon.

JAW BACK, TO: Said of a Sperm Whale when he rolls on his back and fights with his jaws.

JAW-STRAP: A chain sling for hoisting the jaw.

JOIN, TO: To sign ship's articles or papers.

JUNK: The wedge-shaped lower half of the Sperm Whale's forehead, which is above the skull and white horse. It is about equally composed of white horse (meat) and oily matter, both oil and spermaceti. There is no blood in this section, which accounts for the whiteness of the meat, and consequently the name white horse. See Case, and White Horse.

KEEP YOUR EYE PEELED *or* SKINNED: Keep a sharp lookout.

KICKING-STRAP: A rope across the top of the clumsy cleat and fast at each end, under which the whaleline is led. It prevents the whaleline from sweeping aft if it should chance to jump from the chocks.

KILLER WHALE: A variety of the dolphin family, the enemy of Baleen Whales.

KNOCK DOWN THE TRY-WORKS, *or* CASK: To take apart or remove.

KNUCKLE-JOINT: The joint of the whale's flipper which connects with the shoulder blade.

KRENG: The stripped carcass of a whale. Term used in the Greenland Fishery.

LANCE: Instrument for killing whales. See Hand-Lance.

LANCE-LINE: A light line, twelve- or fifteen-thread stuff, with which to recover a lance after it has been tossed.

LAND SHARK: Common name for a sailor's outfitter, applied because of his supposed rapacity.

LANE: A narrow defile in the ice, through which a ship may sail.

LANTERN-KEG: A keg about two feet long, shaped like a truncated cone with a base

A GLOSSARY OF WHALING TERMS

about twelve inches and a head about six inches across. Contains lantern, candle, flint and steel, matches, tobacco, and hard bread; to be used in an emergency.

LARBOARD: The port side of a ship. A term obsolete except aboard a whaler. It is not applied to the rigging or used in navigation. Generally it refers to the position of some article of whaling gear.

LARBOARD BOAT: The boat at the port quarter of a whaler.

LASH-RAIL: A strong rail bolted along the inside of the bulwarks to which the junk, casks, and other deck hamper are lashed. Peculiar to whalers.

LAY: A whaleman's proportionate share of the earnings of a voyage.

LAY THE BOAT ON, TO: To direct it with a single sweep of the steering-oar into the most advantageous position for darting, generally at the right side at right angles, with the bow opposite the flipper.

LAY THE BOAT OFF, TO: To direct the boat out of danger.

LAY THE BOAT AROUND, TO: To turn the boat around by means of the steering-oar.

LEAD (in the ice): A crooked lane, or one that is not clearly open.

LEAN, TO: To cut off any meat which may remain attached to the blubber.

LEANING-KNIFE: Knife employed in leaning.

LEANING-SPADE: Spade employed in leaning; has a very wide blade.

LEVELER PLANE: An arc-shaped tool for planing the heads of casks.

LIFE, THE: The vulnerable spot in a whale, generally the lungs.

LINE: See Whaleline.

LINE-TUB: A large shallow tub in which the line is carefully stowed in successive Flemish coils.

LINESMAN: A seventh man sometimes carried in a British boat during the early nineteenth century, whose duty it was to coil line when not pulling at his oar.

LIPPER: An oblong piece of blubber with a slotted finger grip, used to squeegee the deck after cutting-in.

LIPPER THE DECKS, TO: To scrape the decks with a lipper.

LIVE IRONS: The first and second harpoons which rest, ready to hand, in the boat crotch.

LOBSCOUSE: Salt beef and hard bread hash, sometimes vegetables are added.

LOBTAIL, TO: The act of the Sperm Whale in violently beating the surface of the water with his flukes.

LOGGERHEAD: The projecting timber in the stern of a whaleboat around which the whaleline is snubbed.

LONE BULL: An aged outcast Sperm Whale. He appears to bear about the same relation to a pod of whales that a rogue elephant bears to the elephant herd.

LONG JOINTER: A great plane well over six feet long used in cooperage for beveling and trueing the edges of staves. The front end is supported on a horse and the back end rests against a solid object. The plane is face up and stationary. The cooper thrusts a stave down the face of it.

LONG LAY: The cabin-boy's lay, about one two-hundredth or one two-hundred-twenty-fifth of the voyage. See Short Lay.

LOOKOUT: The masthead sentinel, whose chief duty it is to discover whales. Also the night watch at the forecastle, who reports anything of moment.

Loose Boat: A boat which is not fast to a whale.
Loose Irons: Irons without lines attached. Sometimes used to weaken a whale. An old practice was to dart them into the "small" or root of the tail, which served to slow the creature down.
Loose Whale: A whale with harpoons embedded and lines trailing, but no boats fast; one that has broken away.
Lower away: The order to lower boats for whales.
Make, to: To see or discover anything, generally from a distance; as, "We made No-Man's-Land."
Make a Passage, to: To pass from one whaling ground to another with all sail set.
Making-Iron: A grooved caulking-iron used in finishing off a seam.
Making-Off: Skinning, mincing, and stowing blubber in casks; a name and process of the Greenland Fishery.
Marlinespike: A long, tapered, pointed iron, used to open strands in splicing rope. See Fid, and Pricker.
Masthead, the: The lookout, whose task it is to sight whales. Also his station at the topgallant crosstrees.
Masthead Lookout: Same as above. See Lookout.
Mate, to: To cruise together and divide the catch, said of two whalers. Ships may mate when gamming, or they may not. One is a social arrangement, the other is a business arrangement.
Melon: The case of a blackfish. See Blackfish.
Midship Oar: The waist oar of a whaleboat, the middle or third and longest oar, usually eighteen feet. Generally the term applies to the man who pulls the oar. He sits on the port side, and his rowlock is on the starboard gunwale.
Mik: The Greenland Fishery name for a boat-crotch.
Mill, to: To turn, said of a whale when he makes any considerable change in the direction of his course. More specifically, it means to turn while stationary.
Mince, to: To slice blubber into books.
Mincing-Horse: A plank with a forked end and guiding pegs to hold the horse-pieces while they are minced.
Mincing-Knife: A knife about thirty inches long with a handle at each end, which serves to mince the blubber into thin slices.
Mincing-Tub: The tub which supports the mincing-horse and receives the "books."
Mixed Voyage: One in which whaling was combined with some other venture or ventures; such as codfishing, seal, walrus, bear, and sea-elephant hunting, or trade and barter for hides and pelts.
Monkey Jacket: A short coat. Nothing "long-tailed" was allowed in the boats on account of danger from fouling line.
Monkey Rope: A rope which was either knotted or belted around a man who was sent down on a whale overside for various purposes connected with cutting-in.
Morse: Early English name for walrus, used in the Greenland Fishery.
Mount, a Harpoon or Lance, to: To secure the implement to its pole. The British Greenland Fishery term was "to span in" a harpoon.
Mux: To botch a job.

A GLOSSARY OF WHALING TERMS

Nantucket Sleighride: A ride in a "fast" boat behind a "gallied" whale.
Nib, *or* **Nib End:** The tip of a Baleen Whale's snout.
Nipped: To be pinched in the ice, said of a whaler.
Nipper: A quilted piece of canvas eight or nine inches square which protects the boat-steerer's hand while throwing a turn of the line on or off the loggerhead. A glove or mitten might foul and carry away the hand.
Nisket: The after opening of the alimentary canal of a whale.
Noddle End: The front upper part of the Sperm Whale's snout where the spout hole is located.
Oakum: Old rope picked apart and used for caulking seams.
Oar Apeak, *or* **Peaked:** When the boat is fast, the handle of each oar is slipped into a hole in a cleat fast to the ceiling opposite the rowlock. This "peaks" the oars up at an angle of about twenty degrees, sufficient to clear rough water. The angle of the looms guides the whaleline down the center of the boat.
Off Soundings: Water too deep for a hand lead.
Old Man: The captain. The common way of referring to the captain, but never employed in addressing him.
One: If an officer requires hands, he calls out, "Come here, One!" or "Two!" or "Three!" according to his needs.
Outfit: The equipment for a voyage, either a sailor's or a ship's.
Outrigger Cutting-stage: The modern three-plank staging from which a whale was cut-in.
Pack: Large body of drift ice.
Palm, Sailor's: A leather harness around the hand, with a thimble surface at the root of the thumb: for heavy sewing.
Pan-Bone: The large flat slabs of white bone from the jaws of the Sperm Whale.
Pans, Jaw: Same as Pan-Bone.
Parmaceti: Old name for spermaceti.
Part, to: To break, as a whaleline.
Part, a: A share in a voyage, also called a "piece."
Pay, to: To pour tar or pitch into a seam after caulking.
Peak, to: To stick the handles of the oars in the peak cleats.
Peak Cleat: The cleat which holds the oak apeak. See Oar.
Pick-Haak: Blubber pike and gaff combined, similar to a boathook. Used in Greenland Fishery.
Piece, a: (1) Old term for a part or a share in a whaleship.
(2) A common abbreviation for blanket-piece.
Piggin: A small bucket-like vessel for bailing a boat. One stave is extended to form a handle or ear.
Pike: See Blubber Pike.
Pitch, to: To dart an iron, or a fluke-spade.
Pitch-Pole, to: (1) To dart an iron a long distance by tossing it upward and allowing it to describe a considerable arc before striking.
(2) Said of a whale when he stands vertically with his head out of water, bobbing up and down.

PITCHING: The motion of a whale when, after spouting, his head settles and the hump emerges from the water.

PLUM DUFF:
    1 lb. flour
    1 teaspoonful soda
    2 teaspoonfuls cream of tartar
    2 oz. dripping
    pinch of salt
    6 oz. raisins
    4 oz. sugar.

Sift the flour, soda, cream of tartar, and salt, and add the dripping. Stone the raisins and add the sugar. Mix all together with water. Make into balls and boil for four hours or steam five hours. If allowed, serve with sweet sauce.

This is the plum duff of recent years. Previous to this, potash was used instead of the soda and cream of tartar.

PLUM-PUDDING VOYAGE: A name for a short or 'tween seasons voyage. Also a name applied by the New Bedford whalemen to the short voyages of the Provincetown whalers, the implication being that a Provincetown voyage was a mere picnic.

POD: The common name for a school, herd, or shoal of whales.

POKE: Blackfish or sealskin poke. A skin, bladder, or stomach which is inflated and used as a drug.

POLE: The shaft of a harpoon, lance, or spade. The harpoon pole is of hard wood with the bark still on, as this gives a better hand-hold.

PORPOISE-IRON: An old name for the toggle-iron, which was originally called "Temple's Gig."

POTS AND PANS: The two tin dishes from which the foremast hand eats and drinks.

PREVENTER BOAT-STEERER: A substitute boat-steerer. The foremast hand who is next in line for promotion. He pulls bow oar.

PREVENTER PIN RAILS: Pin rails in the main rigging above the sheer poles to which running rigging is belayed while cutting-in.

PRICKER: A small wooden-handled pointed instrument for splicing and knotting small rope. See Marlinespike, and Fid.

PUMPS: Worn by whalemen in good weather, but kicked from the feet before lowering.

PUT THE BOAT ON A WHALE: See Lay the Boat on.

RAISE WHALES, TO: To discover them; to sight and announce them.

RASING-IRON: A tool for cleaning a seam preparatory to recaulking.

REAMER: Instrument for enlarging and tapering bung-holes.

RECRUITS: (1) Fresh provisions and supplies.
    (2) Various articles taken for purposes of trade, required when replenishing provisions at out-of-the-way ports — calico, beads, powder, tobacco, axes, soap, etc.

RED FLAG: See Flag.

REEM, TO: To wedge seams open with a reeming-iron, so that oakum may be more easily admitted, when caulking.

A GLOSSARY OF WHALING TERMS

REEMING-BEETLE: The caulker's largest mallet.

REEMING-IRON: Used by a caulker for opening seams.

RIDERS: The upper tier casks of the lower hold.

RIDGE: A whale's back between the hump or fin and the small. On the Sperm Whale, beginning at the large dorsal hump, the ridge presents a succession of smaller protuberances diminishing in size as they approach the small.

RIGGER'S BELT: A strong belt which suspends a marlinespike, a grease-horn, and a sheath-knife. The knife is "square-pointed" — that is to say, it has no point.

RIGGER'S HORN: A docked horn containing grease, worn on a rigger's belt.

RIGHT WHALE: The Whalebone Whale of temperate waters; the whale sought by the Yankees in the early shore fishery. He has no hump and is smooth-bellied.

RIPPLE: Disturbance made on the surface when a whale swims just below, without rising, or before rising.

RISING: 1. Motion of a whale when the hump settles and the head lifts out of the water preparatory to spouting. 2. The term applied to each reappearance of the whale at the surface; as, "We'll get fast this rising."

ROLL, TO: To rock from side to side in order to add scope to the snap of the jaw when fighting. Peculiar to Sperm Whales.

ROLLING WHALE: See Roll.

ROUNDING OUT, or ROUNDING: Arching the back nearly at right angles preparatory to turning flukes and sounding.

ROUND-TO, TO: To come to a full stop; said of whales.

RUGGED: Rough or boisterous. Said of the weather or the sea.

RUN, TO: A whale runs, or sounds, settles, bolts, breaches, or rounds-to, but he seldom "swims."

RUSH A CASK, TO: To insert rushes between the staves as a sort of caulking.

RUSHES: Used in flagging casks.

RYER or WRYER: A long narrow cask six or eight inches across the head, made for stowage in the cuntlines between larger casks. It is probable that the name is a corruption of Rider (which see).

SAG HARBOR IRON: A toggle-iron somewhat smaller than that used by the whalemen of other ports.

SALT HORSE: The common name aboard ship for salt beef: similar to corned beef but more briny. It contains a liberal amount of saltpetre, and has to be soaked in the fresh water of the harness cask for about two weeks before it is edible.

SALT JUNK: Another name for the above.

SAVEALLS: Scoops to recover oil and spermaceti from the deck.

SAVE, TO: This is an inclusive term for cutting-in and trying-out; as, "We saved our whale."

SCARF: The line or score around a whale made by the spades in cutting-in.

SCHOOL: A pod of whales.

SCHOOLMASTER: A bull whale much larger than the rest of the school.

SCOOPING: Descriptive name for the Right Whale's method of feeding. He rushes through a field of brit with his mouth distended, gathering great quantities at a single scoop.

Scrag Whale: A name formerly used to designate a Baleen Whale of poor quality — either a thin Right Whale or a Fin Whale, it is not certain which.

Scrap: Blubber from which the oil has been tried. Scrap is the common fuel of the try-works.

Scrap Hopper: A bin for scrap beside the try-pots.

Scriber: A tool for marking casks. It is a compass-like instrument, and the rotating leg makes a circular incision. Straight lines are made with a gouge attachment on the side.

Scrimshaw: The art of the whaleman. Pictorial, ornamental and useful things made of the bone and teeth of the Sperm Whale. Other forms of the word are: Scrimshandy, Scrimshander, Skimshander, Skrimshander, Scrimshonter.

Scrimshaw, to: To practice the Art of Scrimshaw. The usual tools are a saw, a knife, and a file.

Scull, to: To work the steering-oar tailwise to assist in propelling the whaleboat.

Sea-Pig: Porpoise.

Serving-Board: A flat piece of wood with notched edges which answers the same purpose as a serving-mallet, except that it is usually employed for small jobs.

Serving-Mallet: A cylindrical piece of wood with a groove on one side, to fit against a stay, and a handle at the other side, for serving or covering the rigging. It is revolved around a stay, winding and tightening the yarn as it goes.

Set, to get a: To get an opportunity to lance.

Set up, to: To assemble a cask.

Settle, to: To sink without turning flukes or making other observable movement. Peculiar to the Right Whale, although sometimes reported of the Bowhead and Sperm.

Settle a Voyage, to: To divide the proceeds between the owners and the crew.

Shank: Of a harpoon, lance, or spade. The part between the head and the socket.

Share, a (of or in a voyage or ship): Usually an eighth, sixteenth, twenty-fifth, thirty-second, or a sixty-fourth. Also called a "piece" or a "part."

Shark: See Land Shark.

Sheathing, Bottom and Deck: Seven-eighths-inch pine boarding is added to save the structural parts of a ship from wear and deterioration. Copper sheathing is nailed outside the wooden sheathing on the bottom.

Shipkeeper, the: Usually the cooper, who acts as sailing master while the boats are away.

Shipkeepers: Men left to man the ship while boats are away.

Shoal: A school or pod, generally applied to blackfish.

Shooks: Staves, baled for easy stowage.

Short Lay: The captain's lay, or proportionate earnings in a whaling voyage, one-eighth to one-eighteenth. See Long Lay.

Short Warp: (1) The line bent to the second iron. It is attached to the main line with a bowline knot, which runs up on the main line if the iron is fast, or drifts back to the boat if the iron is loose.

(2) The short rope which, in the early New England shore fishery, connected the harpoon and drug.

A GLOSSARY OF WHALING TERMS

Single-Flued Iron: An early hook-shaped harpoon. See illustration.
Size Bone: Whalebone six feet or over in length.
Skeeman: An officer on a British Greenland ship who directed the stowing of the blubber.
Skid Beams: A superstructure for storing spare boats.
Skids: Two thick parallel planks reaching to the gangway from the wharf upon which casks are rolled.
Skimmer: A long-handled copper instrument for removing small bits of scrap from the surface of the boiling oil in the try-pots.
Skipper: "The Old Man"; master or captain of a whaler.
Slack Blubber: The Right Whale when attacked has a habit of sagging or hollowing his back. This leaves a slack, wrinkled section where the blubber does not offer sufficient resistance to be easily penetrated.
Slick: The smooth and oily spot left by a whale in sounding.
Slide Boards: Bent strips on a ship's side, to fend boats in hoisting.
Slop-Chest: The ship's store of ready-made clothes, knives, tobacco, etc., which is drawn upon by the sailor.
Slops: Sailors' garments.
Slumgullion: Refuse of the blubber.
Slush-Tub: A tub in which the fat and grease of the galley is saved. The cook and the ship divide equally whatever this brings when settling a voyage.
Small: The slender part of the whale's after-body where it joins the flukes.
Smooth-Bellied Whales: Sperm, Right, and Bowhead Whales.
Snub, to: To check the whaleline by taking turns around the loggerhead.
So! An order to stop or stay temporarily.
Socket: A hollow cone in a harpoon, lance, or spade, at the base of the shank, into which the pointed pole is fitted.
Soft Bellied Whales: Inclusive name for Humpback, Sulphur-Bottom, and Finback Whales.
Sound, to: To turn flukes and start for the bottom, said of a whale.
Spade: See Cutting-Spade.
Span in, to: To mount an iron on its pole, a term of the British Fishery. (Scoresby, 1820.)
Speak, to: To hail a ship or boat.
Spectioneer: The man who had charge of "flenzing" and "making off" in the Greenland Fishery.
Spermaceti: The case or head matter of the Sperm Whale.
Sperm Whale: The great toothed whale of temperate and tropic waters, absolute monarch of the seas, and the only whale that could have swallowed Jonah.
Spiracle: The spout-hole of a whale.
Spout: The moist visible breath of a whale.
Spouting Blood: Blood appears in the spout when the lungs have been lanced, the sign that the whale is going into his flurry, hence, dying.
Spreaders: Sticks put in the whaleboats, while on the cranes, to keep the gunwales from contracting and warping.

Spring, to: In rowing, to pull hard, to raise from the thwart as the stroke is started.
Spring on the After Oar, to: The act of the mate, who pushes against the after oar to help speed the boat.
Spurs: "Creepers" strapped to the instep to give the men a sure footing while walking on the whale. Used in the old Greenland Fishery.
Square Tier, to Stow: Said when the heads of the ground tier casks and the riders are stowed to the same line.
Squid: Cephalopod Octopus, Sepia Cuttlefish; the food of the Sperm Whale.
Starboard Boat: The after boat on the starboard side. In a "four-boat ship" there is no other on that side. In a "five-boat ship" there is also a "starboard bow boat."
Stave: One of the curved side pieces of a cask.
Steerage: The quarters of the boat-steerers. Usually forward of the cabin on the larboard side. Reached through the booby-hatch.
Steering-Oar: A twenty-two to twenty-three-foot oar for steering a whaleboat. It has a peg at right angles to the loom, one foot from the handle end, providing a grip for the left hand.
Steering-Oar Brace *also called* **Steering Brace:** The fulcrum of the whaleboat's steering-oar, located on a sort of bumpkin, projecting over the larboard side, just forward of the stern post.
Stern All! The order to back the whaleboat away from trouble. Given after striking or lancing.
Stern-Sheets: The space between the after thwart and the stern-cuddy where the wet whaleline is coiled while hauling in.
Stirring-Pole: A stick for stirring oil in the try-pots.
Stocks: British name for harpoon and lance-poles.
Stop-Withers: The reverse barbs or beards on a British two-flued iron.
Stove, to: To smash up a whaleboat; as, "He stove his boat"; "The boat was stove"; "It got stove"; "I am stove"; "Who stove you"; "You'll git stove"; "I'm all stove up."
Stove Boat: A boat that has been damaged by a whale. A whaleboat may also be stove in hoisting and lowering.
Strike, to: To get fast, to fasten, to harpoon.
Sucker: A young suckling whale, the common name for a calf.
Suds: Foam on the surface of the water caused by the actions of a whale.
Sulphur-Bottom: Whalemen's name for the Great Rorqual, or Blue Whale.
Sweep Flukes, to: Said when the whale lifts his flukes from water and swings them from side to side in complete arcs "from eye to eye." A practice of Bowhead and Right Whales.
Swivel Gun: A heavy gun for shooting a harpoon; secured to the bow of a boat.
Sword Fish: The Orca or killer whale.
Tackle, to: To close with, to attack a whale.
Tail Feathers: Stern davits or beams, where boats are stowed on small whalers, usually schooners and brigs.
Take, the: The accumulation of a voyage; as, "Our take was 2000 barrels."
Take, to: To capture a whale. The invariable expression; as, "We took two whales."

A GLOSSARY OF WHALING TERMS

TAKE CARE OF, TO: To perform the necessary offices, as, "We took care of the junk."
TAKE THE LINE, TO: To escape with the line attached. Said of a whale.
TELLTALE: A small compass with a card "upside down" which hangs over the captain's bed. A name also applied to the underside of the double-card steering compass.
TEMPLE'S GIG: See Temple Toggle-Iron.
TEMPLE TOGGLE-IRON: The most successful harpoon ever made. Invented by Lewis Temple, a negro whalecraft-maker of New Bedford, in 1848, and at first called "porpoise-iron" and "Temple's gig." It became the universal whale iron, and has never been improved on.
THERE! A word added for emphasis aboard ship. "What are you doing there?" "Get into that boat, there!" "Hello, there!" "On deck, there!" It has no meaning.
THERE SHE BLOWS! This is the common announcement of the discovery of a whale. The call is repeated each time the whale spouts; often the form is varied, as follows:
  Ah blows!
  She blows!
  There she breeches!
  There she whitewaters!
  She blows and breeches!
  There go flukes!
  Blo-o-o-ows!
  A fish! A fish! (English, 1820)
  Awaite pawana! (Natick Indian; said by St. John to have been used by Nantucketers, 1782. Pawana means whale in Natick.)
  Towno! Towno! *or* Town ho! Town ho! The old hail of the Nantucket Shore Fishery. With this cry the Lookout summoned the townspeople when whales were sighted. The shore fishery was abandoned in 1760.
THIEF: A cylindrical pail about ten inches long, and small enough to enter the bunghole of a cask. Its capacity was about one cupful. Used on whalers at the freshwater butt. It is claimed that when water was scarce, it was kept at the mainmast head. A sailor had to climb to get it and return it when he was through.
THOLE MATS: Thrummed mats to muffle the sound of oars in approaching a whale.
THRASH, TO: To roll and lash the water with flukes and flippers.
THRASHER: A species of shark which attacks baleen whales.
THREE: Collective name used in addressing the starboard oars of a whaleboat; as, "Pull Three," "Stern Three." See Two.
THREE-BOAT SHIP: One that lowered three boats.
TOGGLE, BLUBBER: A hardwood pin, several feet long and about six inches in diameter, used in hoisting blubber. It is buttoned through a strop in the blanket-piece which is fast to the cutting-tackle.
TOGGLE-HEAD: The hinged head of a harpoon.
TOGGLE-IRON: A harpoon with a hinged head, which turns at right angles to the shank when pulled. See Temple Toggle-Iron.

TORMENTOR: The two-pronged fork of the galley, with which salt beef is dished.
TOSS FLUKES, TO: To lift flukes from the water.
TOW-IRON: Early American name for the harpoon used in fastening a boat to a whale. The earlier American iron was used with a short warp and drug and did not tow the boat; it was of much lighter construction and had a spike and shoulder instead of a socket at the base of the shank. One of these early irons is preserved in the Hinsdale collection, and is the only one known to exist. They went out of use between 1761 and 1782.
TOW-ROPE: A name formerly applied to the whaleline.
TOWNO! TOWNO! An old cry of the whaleman when he wanted assistance, usually voiced when in trouble ashore. See "There she blows."
TRAIN-OIL: English name for whale-oil.
TRUSS-HOOPS: Temporary hoops of thick hickory employed while setting up a cask.
TRYING-OUT: The process of boiling oil out of the blubber.
TRY-POTS: Huge iron pots set in the try-works forward.
TRY-WORKS: Brick ovens with try-pots for rendering oil and an insulating water-tank beneath. Built on deck just abaft the forehatch.
TUB-OAR: The fourth oar, the one next the after oar. He sits on the starboard side, and his rowlock is on the port gunwale; also his oar.
TUB-OAR CROTCH: The rowlock of the tub-oar. It is double-decked; that is, it has a branch supporting a higher notch for use if the sea is rough. This lifts the oar sufficiently to clear the line-tub.
TUBS: See Line-Tub.
TURN FLUKES, TO: To toss the flukes in the air and dive. This is the almost invariable gesture of both Sperm and Right Whales preliminary to sounding.
'TWEEN SEASONS VOYAGE: A short voyage, similar to Plum-Pudding Voyage.
TWO: Collective name used in addressing the larboard oars of a whaleboat; as, "Pull Two!" "'Vast pulling Two!" See Three.
TWO-BOAT SHIP: One that lowered two boats. Few ships previous to 1820 lowered more.
TWO-FLUED IRON: The old harpoon, the head of which was arrow-shaped.
UP END A CASK, TO: To stand it with the head up.
UPPER TIER: The second tier of casks in the lower hold; also called "riders."
VOYAGE: The amount of a catch; as, "What voyage did you make?"
VOYAGE: The duration of a cruise for whales.
WAIF (English, *wheft*): A small dark flag with a sharpened staff with which to mark a dead whale. Also used as a signal by the whaleboat.
WAIST BOAT: The midship boat on the larboard or port side of a whaleship.
WATER-KEG *or* BREAKER: A small cask of drinking-water carried in a whaleboat.
WHALE, TO: To go whaling; to take a voyage on a whaleship.
WHALEBOAT: The boat in which whalemen pursue whales.
WHALEBONE: Black "bone" from the mouth of the Right and Bowhead Whales: commercial whalebone.
WHALEBONE WHALE: An inclusive name for Right and Bowhead Whales.

A GLOSSARY OF WHALING TERMS

WHALECRAFT: The iron weapons of capture carried in the whaleboat: harpoons, lances, and spades.

WHALELINE: The rope leading from the line-tub to the harpoon, which fastens a boat to the whale. Long fiber, long-laid manila rope about two and one-quarter inches in circumference.

WHALEMAN: One who has served his apprenticeship on a whaler.

WHALE-OIL: The oil of any Black Whale; specifically the oil of the Right Whale; called by the English, "train-oil."

WHALER: A whaleship.

WHALING GROUNDS: The charted areas which whales are known to frequent, their feeding grounds.

A list of well-known whaling grounds, Bowhead, Sperm, and Right
    Archer Grounds 7°–20° S., 84°–90° W.
    Arctic Ocean (north of Behring Strait)
    Australian Grounds
    Bahamas 28°–29° N. to 79° W.
    Behring Sea
    Bermudas
    Brazil Banks
    Callao Grounds
    Camilla Grounds, or
    Commodore Morris Grounds, } 52°–50° N. 21°–24° W.
    Caribbean Sea (off Chagres)
    Carroll Grounds (between St. Helena and Coast of Africa)
    Charleston Grounds 29°–32° N., 74°–77° W.
    Chile, Coast of
    Congo River
    Crozettes (S.E. of Cape Town)
    Cumberland Inlet
    Davis Straits
    Desolation
    Falkland Islands
    False Banks
    Frobisher Bay
    Grand Banks
    Greenland (the east coast)
    Gulf of Guinea
    Gulf of Mexico 28°–29° N., 89°–90° W.
    Gulf of St. Lawrence
    Hatteras, "Off" (along edge of Gulf Stream)
    Hudson Bay
    Iceland
    Indian Ocean
    Japan Sea
    Kodiak Grounds

Labrador
Madagascar
Main Banks
New Zealand
North Atlantic
North Pacific
Northwest Grounds
Nova Zembla
Off-Shore Ground, 3°–10° S., 90°–120° W. (discovered by ship *Globe*, 1818)
Okhotsk Sea
Patagonia, Coast of
River Plate Ground; also called "Off the River"
Sooloo Grounds (Mindora Seas)
South Atlantic
South Pacific
Spitsbergen
Steen Ground, 31°–36° N., 21°–24° W.
Straits of Belle Isle
St. Helena Grounds
The Twenty Twenties, 20° N., 20° W.
Tristans (Tristan d'Acunha Island)
Two Thirty-Sixes, 36° N.–36° W.
Two Forties, 40° N.–40° W.
West Indies
Western Grounds 28°–36° N., 21°–24° W.

WHERE AWAY? The first query made by the officer of the deck when whales are announced from aloft. It refers to the direction of the whales from ship.

WHITE HORSE: The bloodless meat in the forehead of the Sperm Whale. Specifically, the part of the forehead next above the skull. The sailor's name for all meat is "horse." See Salt Horse.

WHITE WATER: The spray and foam caused by a breaching or thrashing whale.

WHITE WATER, TO: To make spray or suds, said of a whale.

WOOD TO BLACKSKIN: When the stem of the whaleboat collides with the side of a whale, it is said to lie "wood to blackskin."

WOODING: "To go wooding" is to collect wood for the ship either on the beach or ashore.

WRYER: See Ryer.

YOU! Common method of address, officer to sailor.

A LIST OF BOOKS CONCERNING WHALES AND WHALING

ALDRICH, HERBERT L. Arctic Alaska and Siberia, or Eight Months with the Arctic Whalemen. New York, 1887.
ALLEN, GLOVER M. Whalebone Whales of New England. Boston, 1916.
ANDREWS, ROY C. Whale Hunting with Gun and Camera. New York, 1916.
BEALE, THOMAS. Natural History of the Sperm Whale. London, 1839.
BEDDARD, F. E. A Book of Whales. New York, 1900.
BENNETT, FREDERICK D. Narrative of a Whaling Voyage Around the Globe. 2 vols. London, 1840.
BROWN, C. EMERSON. Pocket List of Marine Mammals of Eastern Massachusetts. Salem, 1913.
BROWN, JAMES TEMPLEMAN. Whalemen, Vessels, Apparatus and Methods of the Whale Fishery. In Fishery Industries of the United States. 2 vols. Washington, 1887.
BURNS, W. N. A Year with a Whaler. New York, 1913.
CHEEVER, REV. HENRY T. The Whale and His Captors. Glasgow, n.d.
CHURCHILL, J. and A. Collection of Voyages. London, 1744. (Incidental references.)
CLARK, A. HOWARD. History and Present Condition of the Whalefishery. In Fishery Industries of the United States. 2 vols. Washington, 1887.
COLLECTION OF BOOKS, PAMPHLETS, LOG-BOOKS, PICTURES, ETC., illustrating whales and the Whale Fishery, contained in the Free Public Library, New Bedford, 1907; same, 1920.
CONGDON, BENJAMIN T. List of Shipping Owned in the District of New Bedford. New Bedford, 1835.
CRAPO, HENRY H. New Bedford Directory, 1838.
CRÈVECŒUR (J. HECTOR ST. JOHN, pseud.). Letters of an American Farmer. London, 1782.
DAVIS, WILLIAM M. Nimrod of the Sea. New York, 1874.
DOW, GEORGE FRANCIS. Whaleships and Whaling. Salem, 1925. With a Preface by Frank Wood.
FORBES, ALLAN. Catalogue of Special Exhibition of Whaling Pictures, from the Collection of Allan Forbes, Esq., Peabody Museum. Salem, 1919. (Contains reproductions of whaling prints.)
FORBES, B. C. Loss of the *Essex* (Whaler). Cambridge, 1884.
GIFFORD, PARDON B. Manuscript, History of the Bark *Charles W. Morgan*. New Bedford, 1916.
GOODE, G. BROWN. Fishery Industries of the United States. 2 vols. Washington, 1887.
HARRIS, JOHN. Collection of Voyages and Travels. 2 vols. London, 1748. (Incidental references.)
HART, FRANCIS RUSSELL. New England Whale Fisheries (pamphlet). Cambridge, 1924.

HART, J. C. Miriam Coffin. San Francisco, 1872.
HAWES, CHARLES BOARDMAN. Whaling. New York, 1924.
HOLMES, LEWIS. Arctic Whaleman. Boston, 1857.
HOPKINS, W. J. She Blows! Boston, 1922.
HUSSEY AND ROBINSON. A Catalogue of Nantucket Whalers. Nantucket, 1876.
JARDINE, SIR WILLIAM, BART. Whales. Naturalists' Library, Edinburgh, 1837.
JENKINS, J. T. A History of the Whale Fisheries. London, 1921.
JENKINS, CAPTAIN THOMAS H. Bark *Kathleen* Sunk by a Whale. New Bedford, 1902.
LAING, JOHN. An Account of a Voyage to Spitsbergen, containing a Full Description of that Country, etc. With an Account of the Whale Fishery. London, 1815.
LUCAS, FREDERIC A. The Passing of the Sperm Whale (pamphlet). New York, 1908.
MACPHERSON, DAVID. Annals of Commerce. 4 vols. London, 1805. (Incidental references.)
MACY, OBED. History of Nantucket. Boston, 1835.
MARKHAM, ALBERT HASTINGS, CAPT. R. N. Whaling Cruise to Baffin's Bay. London, 1875.
MAURY, M. F. Sailing Directions for Wind and Current Charts. Philadelphia, 1855. Contains whale charts.
MELVILLE, HERMAN. Moby Dick. New York, 1851.
"MEMORANDUMS." The Outfit of a Whaling Voyage. (A Check-list.) New Bedford, n.d. circa 1850.
MURDOCH, W. G. BURN. Modern Whaling and Bear Hunting. London, 1917.
MURPHY, R. C. Forty-Barrel Bull. New York, 1916.
NORDHOFF, CHARLES. Whaling and Fishing. Cincinnati, 1856.
PEASE, Z. W. Catalpa Expedition. New Bedford, 1897.
RACOVITZA, EMILE G. Spouting and Movements of Whales. Washington, 1904.
RICKETSON, DANIEL. History of New Bedford. New Bedford, 1858.
ST. JOHN, J. HECTOR (CRÈVECŒUR). Letters of an American Farmer. (Chapters on Nantucket Whalefishery.) London, 1782.
SCAMMON, CHARLES M. Marine Mammalia, and an Account of the American Whale Fishery. San Francisco, 1874.
SCORESBY, WILLIAM, JR. Arctic Regions and Northern Whale Fishery. 2 vols. Edinburgh, 1820.
SCORESBY, WILLIAM, JR. Voyage to Greenland. Edinburgh, 1823.
SMITH, CHARLES EDWARD. From the Deep of the Sea. London, 1923.
SPEARS, JOHN R. The Story of the New England Whalers. New York, 1910.
STARBUCK, ALEXANDER. History of the American Whale Fishery. Waltham, 1878.
STATE STREET TRUST COMPANY (compilers). Whale Fishery of New England. Boston, 1915.
THOMES, WILLIAM H. The Whaleman's Adventures in the Sandwich Islands and California. Chicago, 1886.
TOWER, WALTER S. A History of the American Whale Fishery. Philadelphia, 1907.
TRUE, FREDERICK W. Aquatic Mammals. Washington, 1884.
TRUE, FREDERICK W. Finback Whales. Washington, 1903.

TRUE, FREDERICK W. Whalebone Whales of the Western North Atlantic. Washington, 1904.
VILLIERS, A. J. Whaling in the Frozen South. Indianapolis, 1925.
WATSON, ARTHUR C. (Editor). Catalogue of the New Bedford Whaling Museum. New Bedford, 1924.
WHALING DIRECTORY OF THE UNITED STATES. New Bedford, 1869.
WILLIAMS, WILLIAM T. Destruction of the Whaling Fleet in the Arctic Ocean in 1877. New Bedford, 1902.

INDEX

A. R. Tucker, the, 54, 55.
Acushnet, Mass., 34.
Adams, John, on the necessity of protecting whalemen, 25.
Africa, the, 47.
Agents, for whalers, nature of their office, their responsibilities, etc., 107 ff.
Alabama, the, whalers destroyed by, 43.
Amaret, the, notable voyage of, 55.
Ambergris, uses of, 83; found only in "sick" whales, 83.
Amelia, British ship, first whaler to enter Pacific, 27, 38.
American Revolution, effect of, on England's position in whaling industry, 25, and on American whaling fleet, 32.
American whale fishery, how begun, 28.
Americans, in British whale fishery, 26, 27.
Andrews, Roy C., *Whale Hunting with Gun and Camera*, 68, 69, 80.
Ann Alexander, the, 48, 82.
Arctic Bowhead fishery, beginning and importance of, 42; and the *Shenandoah*, 43.
Arctic fleet, steam whalers in, 52, 53.
Armida, the, 96.
Atlantic, the, 82.

Baffin's Bay fishery, 23, 32.
Bailing the "case." *See* Case.
Barentz, William, 23.
Barkentine rig, 52.
Barks, superiority of the rig for whalers, 47.
Basques, in early whaling, 23; harponeers and boat-headers, 24.
Beale, Thomas, *Natural History of the Sperm Whale*, 26, 61, 70, 77, 78.
Beaver, the, first American whaler to round Cape Horn, 38, 60.
Bedford, the, of Nantucket, first vessel to fly U.S. flag in a British port, 38, 47.
Bennett, W. W., 114.
"Bermuda Boat." *See* Marconi rig.
Betsey, the, log of, 92, 93, 96.
Biscay, Bay of, first systematic commercial whaling in, 23.
"Blackskin," 14.
Blankley, *A Naval Expositor*, 46.
"Block Island Boat," 45.
Blubber, 14, 19, 70, 71. *See* Try-works.

Blubber-hooks, difficulty in placing, 15, 16.
Boat crews, how chosen, 5.
Boat-davits, development of, by Yankees, 53, 54; material of, 54.
Boat-headers, 6.
Boat-steerers, 4, 6.
Boats, development of hoisting tackle of, 53; number of, carried by whalers, 54. *See* Whaleboats.
Bomb-lance, 87, 88.
Boston, whaling industry in, 40, 41; one of the last ports to fit out a whaler, 41; mentioned, 31.
Boston Tea Party, 38.
"Boy," on the *Sunbeam*, 7, 9.
Brand, C. C., 87.
Bravas, 108.
Brazil coast, 32.
"Breaching," 81.
Bridgeport, Conn., 40.
Brigs, as whalers, 47.
Brigs, hermaphrodite, as whalers, 52.
Bristol, R.I., 40.
Brown, J. Templeman, in *Fishery Industries of the U.S.*, 58, 60, 61, 63, 77, 78, 100, 112, 115.
Browne, J. Ross, *Etchings of a Whaling Cruise*, 92, 105, 106.
Bullen, F. T., *Cruise of the Cachalot*, 105, 106.
Bulwarks, and their substitute, 48.
By Chance, the, 112.

Canoe, American whaleboat modelled after, 59, 60.
Cape Cod, 28.
Cape Codders, first sporadic attempts at whaling by, 29.
Cape Verde Islands, 32.
Cape Verde islanders, as whalemen, 108.
Captain, the, quarters of, 51; nature of his job, 101.
"Case," bailing the, 18, 19.
Casks, importance of, 97, 98.
Charles W. Morgan, the, last of the New Bedford fleet, 49, 117.
Charleston, S.C., and the Stone Fleet, 42, 43.
Chebec, the, 44.
Cheever, Henry T., *The Whale and his Captors*, 112.
Churchill, J. and A., *Voyages*, 57, 85, 95.

INDEX

Civil War in U.S., deals whaling its first serious blow, 42, 43, 100.
Clark, A. Howard, in *Fishery Industries of the U.S.*, 25, 27, 91.
Clark, Capt. Arthur H., *History of Yachting*, 44, 47.
Clinker boats, 59, 61.
Clinton, Sir Henry, sends expedition against New Bedford, 38.
Clothing of whalemen, quality and price of, 39, 108.
Coldspring Harbor, L.I., 40.
Colonists, English, in America, and the whale fishery, 24.
Commodore Morris, the, 82.
Competition between boats, effect of, 110..
Cook, the, of the *Sunbeam*, 7, 8, 10, 17, 19, 20.
"Coolers," 52.
Cooper, the, of the *Sunbeam*, 2, 3, 4, 7, 9, 16, 17, 20.
Cooper, the, an important person on a whaler, 97, 98.
Crapo, W. W., 104.
Crews, mostly native Americans, 100; furnished by agents, 108; importance of morale of, 109, 110. *See* Boat crews.
Crossjack, 49.
Crow's nest, not used on sperm whalers, 49.
Cunningham, Patrick, 88.
Currier and Ives, prints, 61, 62.
Cutting-in, 14 ff., 95, 96, 97.
Cutting-spades, 94.
Cutting stage, etc., 50, 51.

Danes, in early whaling, 24.
Darting-gun, 88.
Dartmouth, the, first vessel launched at New Bedford, 38; concerned in Boston Tea Party, 38.
Dartmouth, Mass., 34, 38, 40, 92, 96.
Davis, W. M., *Nimrod of the Sea*, 70, 77, 80.
Davis Straits fishery, 23, 24, 25, 32.
Deborah Gifford, the, 54.
Desdemona, the, 73.
Desertions from whalers, 104, 105.
Dexter, Elisha, *Narrative of the Loss of the William and Joseph*, 61.
"Donkey's Breakfast, A," 54.
Dorchester, Mass., 40.
Douglass, William, *Summary*, etc., 91.
"Drudge" ("Drug"), the, 90, 91, 92.
Dudley, Paul, *An Essay*, etc., 60, 91.
Dunkirk, a whaling center, 32.
Durfee, James, 86.
Dutch, the, in early Spitsbergen fishery, 23, 24; first voyages of, to Davis Straits, 24; passing of, as whalers, 26, 27.

Dutch ancestry of early American whalers, 44, 45.

Eastham, Mass., 96.
Easthampton, L.I., 30.
Edgartown, Mass., 39, 40.
Edge, Thomas, 23, 85, 89.
Eleanor B. Conwell, the, 19, 20.
England, vicissitudes of whaling industry in, 25; her diminishing fleet after 1830, 27; methods of dealing with colonial fisheries, 32.
English, the, first whaling voyage by, 23; in Spitsbergen fishery, 24; enter sperm whale fishery in South Seas, 26; wrest supremacy from the Dutch, 26, 27; first to enter the Pacific, 27. *See* South Seas.
Era, the, last American schooner to carry square topsails, 46.
Eskimos, harpoons of, 86; their method of taking whales, 90.
Essex, the, 82.

Fairhaven, Mass., 1; history of whaling in, indissolubly linked with that of New Bedford, 40, 41; statistics, 40; mentioned, 34, 38.
Falconer, *Dictionary of the Marine*, 46.
Fall River, Mass., 40.
Falmouth, Mass., 34, 40.
Fame, the, 89.
"Fastening" to a whale, 92, 93.
Fluke-spade, 87.
Forecastle, the, 52, 54.
Forecastle mess, 7, 8.
Foyn, Svend, 89.
Frates, third mate of the *Sunbeam*, 5, 7, 11, 17, 21.
French, the, in early whaling, 24.

Gaffs and gaffsails, 44, 45, 53.
Galapagos Islands, whaleman's post-office at, 41.
"Gam," the, 103, 104.
Gas, illuminating, and the whaling industry, 42.
Gascoynes, 23.
General Scott, the, 103.
Germans, the, in early whaling, 24.
"Glip," 80.
Gold rush of 1849, its effect on the Pacific fleet, 42.
Gomes, second mate of the *Sunbeam*, 5, 7, 9, 15, 19, 21.
Greenland, East (Spitsbergen), 23, 24.
Greenland, "Old," 23.
"Greenland Fishery," connotation of the phrase, 23.
Greenport, Long Island, 39, 40.

INDEX

Greyhound, the, 54.
Guinea Coast, 32.

Halifax, whaling center, 32.
Hannibal, the, 60.
Harpoon, origin of, 29; development of, 85 ff.; purpose of its use, 87; significance of late return to early types of, 88; poisoned, 89; explosive, fired by cannon, 89.
Harris, John, *Voyages*, 90, 95.
Hart, Joseph C., *Miriam Coffin*, 113.
"Hen Frigate," 50.
Hero, the, first whaling bark, 47.
Higgins, Captain of the *Sunbeam*, his address on leaving port, 5; 2, 7, 10, 16, 19, 20, 77.
Hold, the, 52.
Holmes's Hole, 34, 35, 40.
Honolulu, most important port in Pacific after the gold rush, 42; whalemen responsible for existence of, 42.
Hudson, N.Y., 40.
Hull, England, last English whaler sails from, 27.
Hussey, Christopher, 30, 31.
Hussey and Robinson, *Catalogue of Nantucket Whalers*, 26.

Iceland, 23.
Indians, American, teach colonists how to take whales, 29; their method, 89, 90; as whalemen, 100.

James Arnold, the, 49, 77.
Jamesport, L.I., 40.
Jenkins, J. T., *History of the Whale Fisheries*, 48, 107.
Jib, the, 45.

Kathleen, the, 82.
Kersey, Joseph, 38.
Ketch, the preferred British rig, 46; now being applied to ground trawlers in Mass., 46, 47.

Lapstreak boats. *See* Clinker boats.
Lateen sails, 44.
"Lay," the crew's share of the earnings, 5; average, in late years, 109.
"Laying on," print, 62.
Lewis, William, 52.
"Lobtailing," 80, 81.
Long Island, earliest records of organized whaling from, 29, 30.
Loper, James, 30.
Louisa, the, 77.
Lucas, Frederic A., *The Passing of the Sperm Whale*, 70, 78, 82, 83, 84.

Luck, importance of, in whaling, 55.
Lugsail rig, 58.
Lynn, Mass., 40.

Macpherson, David, *Annals of Commerce*, 25, 26.
Macy, Obed, *History of Nantucket*, 47, 65, 93, 95, 96.
Malloy, Martin, 77.
Manufactor, the, log of a voyage of, in 1756, 35-37, 92; her cruising ground, 37.
"Marconi rig," 46.
Margaret, the, 43, 117.
Maria, the, 49.
Mariner's Mirror, The, 45.
Marion, Mass., 134.
Markham, A. H., *A Whaling Cruise to Baffin's Bay*, 63, 96.
Mary and Helen, the, first American steam whaler to enter the Arctic, 52, 53.
Masthead duty, 5, 6.
Mate and other officers, status of, 51, 52.
Mather, Richard, 29.
Mattapoisett, Mass., 34, 38.
Melville, Herman, *Moby Dick*, a true picture of whaling, 106, 112, 113.
Merchant seamen as whalemen, 100.
Merchants' Transcript, 119.
Mexico, Gulf of, 32.
Milford Haven, whaling center, 32.
"Mixed Voyage," 39.
Monck, *Account of a ... Voyage to Greenland*, 95.
Montanus, Arnoldus, *America*, 45.
Morning Star, the, 58.
Murdoch, W. G. B., *Modern Whaling and Bear Hunting*, 79.
Murphy, R. C., 79.
Mystic, Conn., 40.

Nantucket, whaling fleet of, captured by English in Revolutionary War, 25; the great center of the world's whaling industry, 29; specialization of, in whaling, accounted for, 30 ff.; beginnings of sperm whale fishery at, 30, 31; method of disposing of oil, 31; remote whaling grounds opened up by, 32; gradual depletion of its fleet, 32, 33; the last whaler crosses the bar in 1869, 33; statistics, 40; rigs of fleet of, 47; mentioned, 25, 26, 28, 35, 38, 39, 41, 55, 60, 65, 66, 89, 93, 96, 99, 100, 102.
Nantucket, people of, universally engaged in whaling, 30, 31; characteristics of, 31; mostly impressed by England during Revolutionary War, 32.
New Bedford, number of whalers sailing from, in

INDEX

1854 and 1904, 2; beginning of whaling in district of, 34; its whaling fleet during the Revolution, 37; the only port north of the Chesapeake not in British hands, 38; revival of whaling in, after the war, 38; builds many of its own ships, 38; first vessel launched at, 38; makes most essentials of the industry, 38, 39, 86; growth of the industry in, after 1818, 39; statistics, 40; fleet of, in 1857, 42; rigs of fleet of, 47; sends first American steamer to the Arctic, 52; size of the vessels, 55; last voyages from, successful, and why, 83, 84; last of her whaling fleet, 117 ff.; whalers outfitted by, on Pacific coast, 118; no longer one of the world's great seaports, 120; mentioned, 57, 58, 60, 86, 99, 101, 103, 106–09. *See* Clark's Point.

New Bedford Historical Society and Whaling Museum, 120.

New Bedford *Morning Mercury*, 120.

New England, abundance of whales on coast of, 29, 30.

New London, Conn., vessels from, open up new whaling grounds, 39, and initiate American seal fishery, 39; statistics, 40, 41; first American steam whaler from, 52; mentioned, 103.

New Suffolk, L.I., 40.

New York City, 40.

New York State, ports of, in whaling industry, 39, 40.

Newark, N.J., 40.

Newburyport, Mass., 40.

Newport, R.I., 40.

Nile, the, makes longest whaling voyage on record, 103.

No Duty on Tea, the brig, 37.

Noah, blacksmith of the *Sunbeam*, 5.

Oak, the, last whaler to cross the bar at Nantucket, 33.

Oars, in British and Dutch, and American whaleboats, 62, 63.

Ocean, the, 49.

Officers of whalers, trained in forecastle and steerage, 100, 101; characteristics of, in general, 105. *See* Mate.

Oil, decline in price of, 43. *See* Sperm oil.

Olga, the, 55.

Orca. *See* Whale, Killer.

Osceola 3rd, the, 77.

Pacific Ocean, *Amelia* the first vessel to take whales in, 27, 38.

Paddock, Ichabod, the real father of Nantucket whaling industry, 30.

Paré, Ambroise, 89.

Petroleum, decline of whaling industry caused by, 2, 42, 43.

Philadelphia, whaling in, 39, 48.

Pierce, Eben, 87, 88.

Pilgrims, in the whaling industry, 99.

Pinkies, origin of, 44.

Pioneer, the, first American steam whaler, 52.

Plymouth, Mass., 40.

Point Belcher, whalers destroyed at, 43.

"Poke," the, 90, 91.

Porpoise, *Sunbeam's* first catch of the cruise, 9.

Portland, Me., 40.

Portsmouth, N.H., 40.

Portuguese, as whalemen, 100.

Poughkeepsie, N.Y., 40.

Providence, R.I., 40.

Provincetown, Mass., whaling statistics, 39, 40; in 1868, 47.

Purchas his Pilgrimage (book), 89.

Quakers, in the whaling industry, 99.

Quarterly Review, 26.

Racovitza, Emile G., *A Summary*, etc., 69, 70, 79, 80.

Ready-to-wear clothes, origin of, 39.

Rebecca, the, first American whaler to fill ship on the Pacific, 38.

Ricketson, Daniel, *History of New Bedford*, 34, 37, 38, 39, 42, 92, 96.

Robinson, Albert, 58.

Robinson, Andrew, 45.

Rochester, Mass., 34, 40.

Rotch, William, 32.

Rousseau, the, 49.

Russell, Joseph, pioneer in whaling industry in New Bedford, 34, 37.

Sabine, Lorenzo, *Fisheries of the American Seas*, 29.

Sag Harbor, Long Island, 39, 40.

St. John, J. Hector, *Letters of an American Farmer*, 31, 47, 93, 94, 111.

St. Lawrence, Gulf of, 32.

Salem, Mass., 40.

Sally Ann, the, 60.

San Francisco, and the gold rush (1849), 42; mentioned, 55, 118.

Sandwich Islands. *See* Honolulu.

Sandwich Islanders, as whalemen, 100.

"Saving" the whale, 95, 96.

Scammon, Charles M., *Marine Mammalia*, etc., 62, 75, 77, 91, 115.

"Scarf," 14.

Schooner, prototype of, 45; invention of, claimed by both England and America, 45; origin of the name, 45, 46; American development of the rig,

INDEX

45, 46; when first used in American whaling, 46, 47; advantages and disadvantages of the rig, 46, 47.
Schooners, three-masted, as whalers, 52.
Scoresby, William, *History and Description of the Arctic Regions*, 24, 25, 26, 27, 61, 62, 67, 68; his books the most authoritative, 67; *Arctic Regions and the Northern Whale Fishery*, 71, 72, 74, 75, 88, 93, 94, 96, 109.
Scotland, in the whaling industry, 25.
Scrimshaw, defined, 111; growth and development of, 111 ff.; origin of the word, 112.
Sea, the, largest American whaler on record, 55.
Seal fishery, 39.
Seamanship, the least part of a whaleman's business, 101, 102.
Semmes, Raphael, 43.
Shallop, the, 46.
Shark, harpooning a, 17.
Shenandoah, the, destroys whalers in the Pacific, 43.
Shockley, Humphrey A., 58.
Shockley, William I., 55, 58, **75**, 77, 78, 101.
Sippican. See Marion.
"Slick," 80.
Sloop, 45, 46; large, 47.
Smalley, boat-steerer on the *Sunbeam*, 5, 9, 18, 21.
Smeerenberg, founded by Dutch, 24.
Smith, mate of the *Sunbeam*, 5, 7, 9, 10, 12, 13, 21, 77, 105.
Somerset, Mass., 40.
South Sea Company, sends whalers to Davis Straits, 24.
South Seas, first English sperm-whaling in, 26.
Sparrowhawk, the, 44.
Spears, John R., *Story of the New England Whalers*, 62, 63.
Spencer (sail), 53.
Sperm oil, quantity brought into New Bedford in 1857, 42; uses of, 82.
Sperm whale. See Whale, Sperm.
Sperm whale fishery, entrance of English into, 26; first attempted by English in South Seas, 26; started in Nantucket, 31.
Sperm whalers, special features of their rig, 49.
Spermaceti, 82.
Spitsbergen, named Greenland, 23.
Spitsbergen whale fishery, 23 ff.; supremacy of Dutch in, 24.
"Splicing the mainbrace," 18, 155.
"Spout" of a whale, what it is, 68, 69, 70.
Starbuck, Alexander, *History of the American Whale Fishery*, 34, 47, 66, 100.
Staysails, 45.

Steam, first use of, as an auxiliary, in whaling vessels, 43, 52.
Steel, David, *Elements ... of Rigging and Seamanship*, 94.
"Steerage," significance of, on a whaler, 52.
Steerage mess, 7.
Steward, the, of the *Sunbeam*, 5, 7, 19, 20.
Stone Fleet, the, 43.
Stonington, Conn., 39, 40.
Sullivan, Thomas, 77.
Sunbeam, the, cruise of the author in, 1–22; her age and condition in 1904, 1, 2; divers incidents of the cruise, 12 ff.; its end, 22; desertions from, 104, 105; mentioned, 55, 62, 88.
Sunbeam, crew of the, difficulty in getting them aboard, 2, 3; personnel of, 5, 19.
Sunday, on a whaler, 8, 9, 10.
Susan, the, 115.
Swivel harpoon gun, used by British, 88.

Taber, John, master of the *Manufactor*, 35–37.
Temple, Lewis, 86.
"Temple's Gig," 87.
Theresa, the, pilot-boat, 4.
Thompson, boat-steerer on the *Sunbeam*, 5, 10, 12, 13, 19, 20, 21.
Topsails, square, 46.
Triangular headsails, 45.
Triton, the, 49.
True, Frederick W., *On Some Photographs*, etc., 69.
Truelove, ship, history of, 48.
Try-works, 50, 51, 96.
Two Brothers, the, 58.

Van Bastalaer, R., *Les Etempes*, etc., 45.
Van Luyken, 57.
Van Oelen, J. A., 53, 85.
Vineyard Haven (Holmes's Hole), Mass., 34, 40.
Voyages, length of, 102, 103.

Wanderer, the, last whaler to sail from New Bedford, wrecked, 43, 117.
War of 1812, 32.
Wareham, Mass., 34, 40.
Warren, R.I., 39, 40, 55.
Washington, the, 57.
Waymouth, *Journal of a Voyage to America*, 29, 89, 90.
"Western Halibut rig." See Ketch.
Western Islands, 32.
Westport, Mass., 34, 40.
Whale, Bowhead, 66, 67, 72, 74, 75, 76, 78, 79, 83.
Whale, Finback, 72, 84, 89.
Whale, Humpback, 65, 72, 75, 84, 89.
Whale, Killer, 76.

INDEX

Whale, Right, on New England coast, 30; 65, 66, 67, 72, 74, 76.
Whale, Sperm and Right, difference between, 9; author's first sight of, 10; a natural-born fighter, 65, 66, 82, 92; method of taking, 92, 93; 65, 66, 69, 70, 72, 73, 74, 75, 76, 77, 78, 79, 80, 81, 83, 84.
Whale, Sulphur-Bottom, 72, 79, 84, 87.
Whale, Whalebone. *See* Whale, Bowhead, and Whale, Right.
Whale lance, 87.
Whale oil, how disposed of, by Nantucketers, 31. *See* Sperm oil.
Whaleboats, described, 5; have always carried sails, 57, 58; various rigs of, 58; use of center-board in, results in increase of sail area, 58; an unrivalled surf-boat, 59; difference between American, and British and Dutch, 59, 61; typical British, 59, 60; American models of, 59, 60; materials of, 61; position of oarsmen in, 61; divers details concerning, 62–64.
Whalebone, diminished use of, 82, 83; 42, 43, 66.
Whaleline, use of, 89, 92, 93; material of, 94.
Whaleman's post-office on Galapagos, 41.
Whaleman's Shipping List, 119, 120.
Whalemen, paid by "lay," not by wages, 5, 107, 108, 109; apparel of, 39; achievements of, 102; qualifications of, 102; effect of tedium of long voyages, 102, 103.
Whalers (vessels), decrease in numbers of, 2; competition between, 42; early American, built on Dutch lines, 44; large sloops as, 47; the best rig for, 47; general character of, 47, 48; average life of, 48; cruising rig of, 48; typical, when evolved, 49; bark-rigged, 49; general description of, 49 ff.; developments of, applied to other vessels, 53; number of boats carried by, 54; hull construction of, 54, 55; value of, in 1800 and 1900, 55; increase in average tonnage of, followed by decrease, 55; source of reputation of, for evil odor, 55, 56; quality and variety of food on, 106, 107. *See* Whaling industry.
Whales, various methods of taking, 29, 89, 90; stranded, salving of, the beginning of the American fishery, 28; methods of trying out, 28, 29, 96; inveterate habits of, 41; particulars concerning, 67 ff.; age of, 73; what constitutes a big one, 76, 77, 78, 79; vision of, 78, 79; supposed to have means of communication at a distance, 80; every part of, now made into marketable products, 83; methods of "saving," 95, 96.
Whaling fleet of the U.S., size and distribution of, in 1839, 40; size of, in 1842, compared to that of the rest of the world, 41; size of, in 1857, 42; destruction of, in the Civil War, 42, 43.
"Whaling for glory," 110.
Whaling grounds, list of, in the order of their discovery, 41; alphabetical list, 145, 146.
Whaling industry, the least changed by modern conditions, of all industries, 1; why dead to-day, 2; boatmen, not seamen, required in, 5; early history of, 23 ff.; rivalry of British and Dutch in, 26, 27; earliest records of, as an organized industry, 29, 30; in New Bedford, 34, 38; in the Pacific, beginning of, 38; extends to divers other ports after the Revolution, 38, 39, 41; history of, largely a record of discovery of new grounds, 41; Golden Age of, 42; improvements in tools and gear of, 42; and the introduction of gas, 42; effect of Civil War on, 42, 43, 100; and petroleum, 42, 43; brief revival of, 43; effect of use of try-works on, 96; carried on chiefly by Quakers in Nantucket and New Bedford, 99, 100.
Wilmington, Del., whaling in, 39, 40.
Windlass, the, change in method of working, 49, 50.
Wing, J. and W. R., owners of the *Sunbeam*, 2.
Winslow, George, 73.
Wiscasset, Me., 40.

A CATALOG OF SELECTED
DOVER BOOKS
IN ALL FIELDS OF INTEREST

A CATALOG OF SELECTED
DOVER BOOKS
IN ALL FIELDS OF INTEREST

DRAWINGS OF REMBRANDT, edited by Seymour Slive. Updated Lippmann, Hofstede de Groot edition, with definitive scholarly apparatus. All portraits, biblical sketches, landscapes, nudes. Oriental figures, classical studies, together with selection of work by followers. 550 illustrations. Total of 630pp. 9⅛ × 12¼. 21485-0, 21486-9 Pa., Two-vol. set $29.90

GHOST AND HORROR STORIES OF AMBROSE BIERCE, Ambrose Bierce. 24 tales vividly imagined, strangely prophetic, and decades ahead of their time in technical skill: "The Damned Thing," "An Inhabitant of Carcosa," "The Eyes of the Panther," "Moxon's Master," and 20 more. 199pp. 5⅜ × 8½. 20767-6 Pa. $4.95

ETHICAL WRITINGS OF MAIMONIDES, Maimonides. Most significant ethical works of great medieval sage, newly translated for utmost precision, readability. Laws Concerning Character Traits, Eight Chapters, more. 192pp. 5⅜ × 8½. 24522-5 Pa. $5.95

THE EXPLORATION OF THE COLORADO RIVER AND ITS CANYONS, J. W. Powell. Full text of Powell's 1,000-mile expedition down the fabled Colorado in 1869. Superb account of terrain, geology, vegetation, Indians, famine, mutiny, treacherous rapids, mighty canyons, during exploration of last unknown part of continental U.S. 400pp. 5⅜ × 8½. 20094-9 Pa. $8.95

HISTORY OF PHILOSOPHY, Julián Marías. Clearest one-volume history on the market. Every major philosopher and dozens of others, to Existentialism and later. 505pp. 5⅜ × 8½. 21739-6 Pa. $9.95

ALL ABOUT LIGHTNING, Martin A. Uman. Highly readable nontechnical survey of nature and causes of lightning, thunderstorms, ball lightning, St. Elmo's Fire, much more. Illustrated. 192pp. 5⅜ × 8½. 25237-X Pa. $5.95

SAILING ALONE AROUND THE WORLD, Captain Joshua Slocum. First man to sail around the world, alone, in small boat. One of great feats of seamanship told in delightful manner. 67 illustrations. 294pp. 5⅜ × 8½. 20326-3 Pa. $4.95

LETTERS AND NOTES ON THE MANNERS, CUSTOMS AND CONDITIONS OF THE NORTH AMERICAN INDIANS, George Catlin. Classic account of life among Plains Indians: ceremonies, hunt, warfare, etc. 312 plates. 572pp. of text. 6⅛ × 9¼. 22118-0, 22119-9, Pa., Two-vol. set $17.90

THE SECRET LIFE OF SALVADOR DALÍ, Salvador Dalí. Outrageous but fascinating autobiography through Dalí's thirties with scores of drawings and sketches and 80 photographs. A must for lovers of 20th-century art. 432pp. 6½ × 9¼. (Available in U.S. only) 27454-3 Pa. $9.95

CATALOG OF DOVER BOOKS

SIR HARRY HOTSPUR OF HUMBLETHWAITE, Anthony Trollope. Incisive, unconventional psychological study of a conflict between a wealthy baronet, his idealistic daughter, and their scapegrace cousin. The 1870 novel in its first inexpensive edition in years. 250pp. 5⅜ × 8½. 24953-0 Pa. $6.95

LASERS AND HOLOGRAPHY, Winston E. Kock. Sound introduction to burgeoning field, expanded (1981) for second edition. Wave patterns, coherence, lasers, diffraction, zone plates, properties of holograms, recent advances. 84 illustrations. 160pp. 5⅜ × 8¼. (Except in United Kingdom) 24041-X Pa. $4.95

INTRODUCTION TO ARTIFICIAL INTELLIGENCE: Second, Enlarged Edition, Philip C. Jackson, Jr. Comprehensive survey of artificial intelligence—the study of how machines (computers) can be made to act intelligently. Includes introductory and advanced material. Extensive notes updating the main text. 132 black-and-white illustrations. 512pp. 5⅜ × 8½. 24864-X Pa. $10.95

HISTORY OF INDIAN AND INDONESIAN ART, Ananda K. Coomaraswamy. Over 400 illustrations illuminate classic study of Indian art from earliest Harappa finds to early 20th century. Provides philosophical, religious and social insights. 304pp. 6⅜ × 9⅜. 25005-9 Pa. $11.95

THE GOLEM, Gustav Meyrink. Most famous supernatural novel in modern European literature, set in Ghetto of Old Prague around 1890. Compelling story of mystical experiences, strange transformations, profound terror. 13 black-and-white illustrations. 224pp. 5⅜ × 8½. 25025-3 Pa. $7.95

PICTORIAL ENCYCLOPEDIA OF HISTORIC ARCHITECTURAL PLANS, DETAILS AND ELEMENTS: With 1,880 Line Drawings of Arches, Domes, Doorways, Facades, Gables, Windows, etc., John Theodore Haneman. Sourcebook of inspiration for architects, designers, others. Bibliography. Captions. 141pp. 9 × 12. 24605-1 Pa. $8.95

BENCHLEY LOST AND FOUND, Robert Benchley. Finest humor from early 30s, about pet peeves, child psychologists, post office and others. Mostly unavailable elsewhere. 73 illustrations by Peter Arno and others. 183pp. 5⅜ × 8½. 22410-4 Pa. $4.95

ERTÉ GRAPHICS, Erté. Collection of striking color graphics: *Seasons, Alphabet, Numerals, Aces* and *Precious Stones*. 50 plates, including 4 on covers. 48pp. 9⅜ × 12¼. 23580-7 Pa. $7.95

THE JOURNAL OF HENRY D. THOREAU, edited by Bradford Torrey, F. H. Allen. Complete reprinting of 14 volumes, 1837–61, over two million words; the sourcebooks for *Walden*, etc. Definitive. All original sketches, plus 75 photographs. 1,804pp. 8½ × 12¼. 20312-3, 20313-1 Cloth., Two-vol. set $130.00

CASTLES: Their Construction and History, Sidney Toy. Traces castle development from ancient roots. Nearly 200 photographs and drawings illustrate moats, keeps, baileys, many other features. Caernarvon, Dover Castles, Hadrian's Wall, Tower of London, dozens more. 256pp. 5⅜ × 8¼. 24898-4 Pa. $7.95

CATALOG OF DOVER BOOKS

DEGAS: An Intimate Portrait, Ambroise Vollard. Charming, anecdotal memoir by famous art dealer of one of the greatest 19th-century French painters. 14 black-and-white illustrations. Introduction by Harold L. Van Doren. 96pp. 5⅜ × 8½. 25131-4 Pa. $4.95

PERSONAL NARRATIVE OF A PILGRIMAGE TO AL-MADINAH AND MECCAH, Richard F. Burton. Great travel classic by remarkably colorful personality. Burton, disguised as a Moroccan, visited sacred shrines of Islam, narrowly escaping death. 47 illustrations. 959pp. 5⅜ × 8½. 21217-3, 21218-1 Pa., Two-vol. set $19.90

PHRASE AND WORD ORIGINS, A. H. Holt. Entertaining, reliable, modern study of more than 1,200 colorful words, phrases, origins and histories. Much unexpected information. 254pp. 5⅜ × 8½. 20758-7 Pa. $5.95

THE RED THUMB MARK, R. Austin Freeman. In this first Dr. Thorndyke case, the great scientific detective draws fascinating conclusions from the nature of a single fingerprint. Exciting story, authentic science. 320pp. 5⅜ × 8½. (Available in U.S. only) 25210-8 Pa. $6.95

AN EGYPTIAN HIEROGLYPHIC DICTIONARY, E. A. Wallis Budge. Monumental work containing about 25,000 words or terms that occur in texts ranging from 3000 B.C. to 600 A.D. Each entry consists of a transliteration of the word, the word in hieroglyphs, and the meaning in English. 1,314pp. 6⅝ × 10. 23615-3, 23616-1 Pa., Two-vol. set $35.90

THE COMPLEAT STRATEGYST: Being a Primer on the Theory of Games of Strategy, J. D. Williams. Highly entertaining classic describes, with many illustrated examples, how to select best strategies in conflict situations. Prefaces. Appendices. xvi + 268pp. 5⅜ × 8½. 25101-2 Pa. $7.95

THE ROAD TO OZ, L. Frank Baum. Dorothy meets the Shaggy Man, little Button-Bright and the Rainbow's beautiful daughter in this delightful trip to the magical Land of Oz. 272pp. 5⅜ × 8. 25208-6 Pa. $5.95

POINT AND LINE TO PLANE, Wassily Kandinsky. Seminal exposition of role of point, line, other elements in nonobjective painting. Essential to understanding 20th-century art. 127 illustrations. 192pp. 6½ × 9¼. 23808-3 Pa. $5.95

LADY ANNA, Anthony Trollope. Moving chronicle of Countess Lovel's bitter struggle to win for herself and daughter Anna their rightful rank and fortune—perhaps at cost of sanity itself. 384pp. 5⅜ × 8½. 24669-8 Pa. $8.95

EGYPTIAN MAGIC, E. A. Wallis Budge. Sums up all that is known about magic in Ancient Egypt: the role of magic in controlling the gods, powerful amulets that warded off evil spirits, scarabs of immortality, use of wax images, formulas and spells, the secret name, much more. 253pp. 5⅜ × 8½. 22681-6 Pa. $4.95

THE DANCE OF SIVA, Ananda Coomaraswamy. Preeminent authority unfolds the vast metaphysic of India: the revelation of her art, conception of the universe, social organization, etc. 27 reproductions of art masterpieces. 192pp. 5⅜ × 8½. 24817-8 Pa. $6.95

CATALOG OF DOVER BOOKS

CHRISTMAS CUSTOMS AND TRADITIONS, Clement A. Miles. Origin, evolution, significance of religious, secular practices. Caroling, gifts, yule logs, much more. Full, scholarly yet fascinating; non-sectarian. 400pp. 5⅜ × 8½.
23354-5 Pa. $7.95

THE HUMAN FIGURE IN MOTION, Eadweard Muybridge. More than 4,500 stopped-action photos, in action series, showing undraped men, women, children jumping, lying down, throwing, sitting, wrestling, carrying, etc. 390pp. 7⅞ × 10⅝.
20204-6 Cloth. $24.95

THE MAN WHO WAS THURSDAY, Gilbert Keith Chesterton. Witty, fast-paced novel about a club of anarchists in turn-of-the-century London. Brilliant social, religious, philosophical speculations. 128pp. 5⅜ × 8½.
25121-7 Pa. $3.95

A CÉZANNE SKETCHBOOK: Figures, Portraits, Landscapes and Still Lifes, Paul Cézanne. Great artist experiments with tonal effects, light, mass, other qualities in over 100 drawings. A revealing view of developing master painter, precursor of Cubism. 102 black-and-white illustrations. 144pp. 8¾ × 6⅜.
24790-2 Pa. $6.95

AN ENCYCLOPEDIA OF BATTLES: Accounts of Over 1,560 Battles from 1479 B.C. to the Present, David Eggenberger. Presents essential details of every major battle in recorded history, from the first battle of Megiddo in 1479 B.C. to Grenada in 1984. List of Battle Maps. New Appendix covering the years 1967–1984. Index. 99 illustrations. 544pp. 6½ × 9¼.
24913-1 Pa. $14.95

AN ETYMOLOGICAL DICTIONARY OF MODERN ENGLISH, Ernest Weekley. Richest, fullest work, by foremost British lexicographer. Detailed word histories. Inexhaustible. Total of 856pp. 6½ × 9¼.
21873-2, 21874-0 Pa., Two-vol. set $19.90

WEBSTER'S AMERICAN MILITARY BIOGRAPHIES, edited by Robert McHenry. Over 1,000 figures who shaped 3 centuries of American military history. Detailed biographies of Nathan Hale, Douglas MacArthur, Mary Hallaren, others. Chronologies of engagements, more. Introduction. Addenda. 1,033 entries in alphabetical order. xi + 548pp. 6½ × 9¼. (Available in U.S. only)
24758-9 Pa. $13.95

LIFE IN ANCIENT EGYPT, Adolf Erman. Detailed older account, with much not in more recent books: domestic life, religion, magic, medicine, commerce, and whatever else needed for complete picture. Many illustrations. 597pp. 5⅜ × 8½.
22632-8 Pa. $9.95

HISTORIC COSTUME IN PICTURES, Braun & Schneider. Over 1,450 costumed figures shown, covering a wide variety of peoples: kings, emperors, nobles, priests, servants, soldiers, scholars, townsfolk, peasants, merchants, courtiers, cavaliers, and more. 256pp. 8⅜ × 11¼.
23150-X Pa. $9.95

THE NOTEBOOKS OF LEONARDO DA VINCI, edited by J. P. Richter. Extracts from manuscripts reveal great genius; on painting, sculpture, anatomy, sciences, geography, etc. Both Italian and English. 186 ms. pages reproduced, plus 500 additional drawings, including studies for *Last Supper*, *Sforza* monument, etc. 860pp. 7⅞ × 10¾.
22572-0, 22573-9 Pa., Two-vol. set $35.90

CATALOG OF DOVER BOOKS

THE ART NOUVEAU STYLE BOOK OF ALPHONSE MUCHA: All 72 Plates from "Documents Decoratifs" in Original Color, Alphonse Mucha. Rare copyright-free design portfolio by high priest of Art Nouveau. Jewelry, wallpaper, stained glass, furniture, figure studies, plant and animal motifs, etc. Only complete one-volume edition. 80pp. 9⅜ × 12¼. 24044-4 Pa. $10.95

ANIMALS: 1,419 Copyright-Free Illustrations of Mammals, Birds, Fish, Insects, Etc., edited by Jim Harter. Clear wood engravings present, in extremely lifelike poses, over 1,000 species of animals. One of the most extensive pictorial sourcebooks of its kind. Captions. Index. 284pp. 9 × 12. 23766-4 Pa. $10.95

OBELISTS FLY HIGH, C. Daly King. Masterpiece of American detective fiction, long out of print, involves murder on a 1935 transcontinental flight—"a very thrilling story"—NY Times. Unabridged and unaltered republication of the edition published by William Collins Sons & Co. Ltd., London, 1935. 288pp. 5⅜ × 8½. (Available in U.S. only) 25036-9 Pa. $5.95

VICTORIAN AND EDWARDIAN FASHION: A Photographic Survey, Alison Gernsheim. First fashion history completely illustrated by contemporary photographs. Full text plus 235 photos, 1840-1914, in which many celebrities appear. 240pp. 6½ × 9¼. 24205-6 Pa. $8.95

THE ART OF THE FRENCH ILLUSTRATED BOOK, 1700-1914, Gordon N. Ray. Over 630 superb book illustrations by Fragonard, Delacroix, Daumier, Doré, Grandville, Manet, Mucha, Steinlen, Toulouse-Lautrec and many others. Preface. Introduction. 633 halftones. Indices of artists, authors & titles, binders and provenances. Appendices. Bibliography. 608pp. 8⅜ × 11¼. 25086-5 Pa. $24.95

THE WONDERFUL WIZARD OF OZ, L. Frank Baum. Facsimile in full color of America's finest children's classic. 143 illustrations by W. W. Denslow. 267pp. 5⅜ × 8½. 20691-2 Pa. $7.95

FOLLOWING THE EQUATOR: A Journey Around the World, Mark Twain. Great writer's 1897 account of circumnavigating the globe by steamship. Ironic humor, keen observations, vivid and fascinating descriptions of exotic places. 197 illustrations. 720pp. 5⅜ × 8½. 26113-1 Pa. $15.95

THE FRIENDLY STARS, Martha Evans Martin & Donald Howard Menzel. Classic text marshalls the stars together in an engaging, nontechnical survey, presenting them as sources of beauty in night sky. 23 illustrations. Foreword. 2 star charts. Index. 147pp. 5⅜ × 8½. 21099-5 Pa. $3.95

FADS AND FALLACIES IN THE NAME OF SCIENCE, Martin Gardner. Fair, witty appraisal of cranks, quacks, and quackeries of science and pseudoscience: hollow earth, Velikovsky, orgone energy, Dianetics, flying saucers, Bridey Murphy, food and medical fads, etc. Revised, expanded In the Name of Science. "A very able and even-tempered presentation."—The New Yorker. 363pp. 5⅜ × 8. 20394-8 Pa. $6.95

ANCIENT EGYPT: Its Culture and History, J. E. Manchip White. From predynastics through Ptolemies: society, history, political structure, religion, daily life, literature, cultural heritage. 48 plates. 217pp. 5⅜ × 8½. 22548-8 Pa. $5.95

CATALOG OF DOVER BOOKS

AMERICAN CLIPPER SHIPS: 1833–1858, Octavius T. Howe & Frederick C. Matthews. Fully-illustrated, encyclopedic review of 352 clipper ships from the period of America's greatest maritime supremacy. Introduction. 109 halftones. 5 black-and-white line illustrations. Index. Total of 928pp. 5⅜ × 8½.
25115-2, 25116-0 Pa., Two-vol. set $21.90

TOWARDS A NEW ARCHITECTURE, Le Corbusier. Pioneering manifesto by great architect, near legendary founder of "International School." Technical and aesthetic theories, views on industry, economics, relation of form to function, "mass-production spirit," much more. Profusely illustrated. Unabridged translation of 13th French edition. Introduction by Frederick Etchells. 320pp. 6⅛ × 9¼. (Available in U.S. only)
25023-7 Pa. $8.95

THE BOOK OF KELLS, edited by Blanche Cirker. Inexpensive collection of 32 full-color, full-page plates from the greatest illuminated manuscript of the Middle Ages, painstakingly reproduced from rare facsimile edition. Publisher's Note. Captions. 32pp. 9⅜ × 12¼. (Available in U.S. only)
24345-1 Pa. $5.95

BEST SCIENCE FICTION STORIES OF H. G. WELLS, H. G. Wells. Full novel *The Invisible Man*, plus 17 short stories: "The Crystal Egg," "Aepyornis Island," "The Strange Orchid," etc. 303pp. 5⅜ × 8½. (Available in U.S. only)
21531-8 Pa. $6.95

AMERICAN SAILING SHIPS: Their Plans and History, Charles G. Davis. Photos, construction details of schooners, frigates, clippers, other sailcraft of 18th to early 20th centuries—plus entertaining discourse on design, rigging, nautical lore, much more. 137 black-and-white illustrations. 240pp. 6⅛ × 9¼.
24658-2 Pa. $6.95

ENTERTAINING MATHEMATICAL PUZZLES, Martin Gardner. Selection of author's favorite conundrums involving arithmetic, money, speed, etc., with lively commentary. Complete solutions. 112pp. 5⅜ × 8½.
25211-6 Pa. $3.95

THE WILL TO BELIEVE, HUMAN IMMORTALITY, William James. Two books bound together. Effect of irrational on logical, and arguments for human immortality. 402pp. 5⅜ × 8½.
20291-7 Pa. $8.95

THE HAUNTED MONASTERY and THE CHINESE MAZE MURDERS, Robert Van Gulik. 2 full novels by Van Gulik continue adventures of Judge Dee and his companions. An evil Taoist monastery, seemingly supernatural events; overgrown topiary maze that hides strange crimes. Set in 7th-century China. 27 illustrations. 328pp. 5⅜ × 8½.
23502-5 Pa. $6.95

CELEBRATED CASES OF JUDGE DEE (DEE GOONG AN), translated by Robert Van Gulik. Authentic 18th-century Chinese detective novel; Dee and associates solve three interlocked cases. Led to Van Gulik's own stories with same characters. Extensive introduction. 9 illustrations. 237pp. 5⅜ × 8½.
23337-5 Pa. $5.95

Prices subject to change without notice.

Available at your book dealer or write for free catalog to Dept. GI, Dover Publications, Inc., 31 East 2nd St., Mineola, N.Y. 11501. Dover publishes more than 400 books each year on science, elementary and advanced mathematics, biology, music, art, literary history, social sciences and other areas.